CHEMISTRY

Focus
Review Guide
for AP® Chemistry

 12e

CHANG | GOLDSBY

Mc Graw Hill Education

AP®, Advanced Placement®, and Advanced Placement Program® are trademarks registered and/or owned by the College Board, which was not involved in the production of, and does not endorse, these products.

About the AP Consultant

Todd Abronowitz was the lead AP Chemistry teacher for Dallas ISD and currently teaches at Parish Episcopal School in Dallas, TX. He has taught AP Chemistry since 1993 and has been a College Board consultant for AP and Pre-AP Chemistry since 2003. He has received numerous local, state and national awards for his teaching including the 2000 Texas Chemistry Teacher of the Year, 2003 Radio Shack National Teacher Award Winner, and the 2012 National Math and Science Initiative All American Teacher of the Year. He has taught in numerous settings in the states of Michigan and Texas: rural, suburban, and urban public school districts and parochial schools. He has also taught at the college level, most recently for Rice University.

MHEonline.com

Send all inquiries to:
McGraw-Hill Education
8787 Orion Place
Columbus, OH 43240

ISBN: 978-0-07-672803-9
MHID: 0-07-672803-X

Printed in the United States of America.

2 3 4 5 6 7 QVS 20 19 18 17 16

Table of Contents

Chapter	Section	AP Correlation
Chapter 1 Chemistry: The Study of Change pp. 1–9	1.1	Prerequisite Knowledge
	1.2	Prerequisite Knowledge
	1.3	Prerequisite Knowledge
	1.4	1.A.1, 2.A.3
	1.5	1.A.1, 2.A.1, 2.A.2
	1.6	Extending Knowledge
	1.7	Extending Knowledge
	1.8	Extending Knowledge
	1.9	Extending Knowledge
	1.10	Extending Knowledge
Chapter 2 Atoms, Molecules, and Ions pp. 10–17	2.1	1.A.1, 1.D.1
	2.2	1.B.1, 1.D.1
	2.3	1.D.1, 1.D.2
	2.4	1.C.1
	2.5	1.A.1
	2.6	1.A.1, 2.C.2
	2.7	Prerequisite Knowledge
	2.8	Extending Knowledge
Chapter 3 Mass Relationships in Chemical Reactions pp. 18–30	3.1	1.A.3
	3.2	1.A.3
	3.3	1.A.3
	3.4	1.D.2
	3.5	1.A.1, 1.A.2, 1.A.3
	3.6	1.A.2, 1.A.3
	3.7	1.A.1, 1.E.1, 1.E.2, 3.A.1, 3.C.1
	3.8	1.A.2, 1.A.3, 1.E.1, 1.E.2, 3.A.1, 3.A.2
	3.9	1.A.3, 3.A.1, 3.A.2
	3.10	1.A.1, 1.A.3, 3.A.1, 3.A.2

Chapter	Section	AP Correlation
Chapter 4 Reactions in Aqueous Solutions pp. 31–42	4.1	2.A.1, 2.A.3, 2.D.1, 6.A.1
	4.2	3.A.1
	4.3	1.E.1, 3.A.1, 3.B.2
	4.4	1.E.1, 3.A.1, 3.B.1, 3.B.3
	4.5	2.A.3
	4.6	1.A.2, 1.E.2, 3.A.1, 3.A.2
	4.7	3.A.2, 3.B.2
	4.8	1.A.2, 1.E.2, 3.A.2, 3.B.3
Chapter 5 Gases pp. 43–54	5.1	2.B.3
	5.2	2.A.2
	5.3	1.E.1, 2.A.2
	5.4	2.A.2
	5.5	2.A.1, 2.A.2, 3.A.2
	5.6	2.A.2
	5.7	2.A.2, 5.A.1
	5.8	2.A.2
Chapter 6 Thermochemistry pp. 55–65	6.1	5.B.1, 5.B.2, 5.C.1, 5.D.1
	6.2	3.C.2, 5.B.1, 5.B.2, 5.B.3, 5.C.2
	6.3	3.C.2, 5.B.1, 5.B.2, 5.B.3
	6.4	1.E.2, 3.C.2, 5.B.1, 5.B.2, 5.B.3
	6.5	5.B.1, 5.B.2, 5.B.3, 5.B.4
	6.6	5.C.2
	6.7	5.B.3, 5.C.2
Chapter 7 Quantum Theory and the Electronic Structure of Atoms pp. 66–79	7.1	1.C.2
	7.2	1.B.1, 1.D.3
	7.3	1.C.2, 1.D.3
	7.4	1.C.2, 1.D.3
	7.5	1.C.2, 1.D.1
	7.6	1.B.2, 1.C.2
	7.7	1.B.2
	7.8	1.B.2
	7.9	1.B.2, 1.C.1

Chapter	Section	AP Correlation
Chapter 8 Periodic Relationships Among the Elements pp. 80–87	8.1	Prerequisite Knowledge
	8.2	1.C.1
	8.3	1.C.1
	8.4	1.C.1
	8.5	1.C.1
	8.6	1.C.1
Chapter 9 Chemical Bonding I: Basic Concepts pp. 88–101	9.1	1.C.1, 2.C.1, 2.C.2
	9.2	1.C.2, 2.C.2
	9.3	1.B.1, 2.C.2
	9.4	2.C.1, 2.D.1, 2.D.2
	9.5	1.C.1, 2.C.1
	9.6	1.C.1, 2.C.1
	9.7	Prerequisite Knowledge
	9.8	2.C.4
	9.9	2.C.4
	9.10	3.C.2, 5.C.1, 5.C.2
Chapter 10 Chemical Bonding II: Molecular Geometry and Hybridization of Atomic Orbitals pp. 102–112	10.1	2.C.4
	10.2	2.C.1
	10.3	2.C.4
	10.4	2.C.4
	10.5	2.C.4
	10.6	Extending Knowledge
	10.7	Extending Knowledge
	10.8	2.C.4
Chapter 11 Intermolecular Forces and Liquids and Solids pp. 113–122	11.1	2.A.1
	11.2	2.B.1, 2.B.2, 5.D.1
	11.3	2.A.1
	11.4	Extending Knowledge
	11.5	Extending Knowledge
	11.6	2.C.3, 2.D.1, 2.D.2, 2.D.3
	11.7	Extending Knowledge
	11.8	2.A.2, 5.B.3, 5.D.1, 6.A.1
	11.9	Extending Knowledge

Chapter	Section	AP Correlation
Chapter 12 Physical Properties of Solutions pp. 123–132	12.1	2.A.3
	12.2	2.A.3, 2.B.3, 5.E.1
	12.3	2.A.3
	12.4	Extending Knowledge
	12.5	Extending Knowledge
	12.6	Prerequisite Knowledge
	12.7	Prerequisite Knowledge
	12.8	Extending Knowledge
Chapter 13 Chemical Kinetics pp. 133–141	13.1	4.A.1, 4.A.3
	13.2	4.A.1, 4.A.2, 4.A.3
	13.3	4.A.1, 4.A.2
	13.4	4.B.1, 4.B.2, 4.B.3
	13.5	4.C.1, 4.C.2, 4.C.3
	13.6	4.B.3, 4.D.1, 4.D.2
Chapter 14 Chemical Equilibrium pp. 142–148	14.1	6.A.4
	14.2	6.A.1, 6.A.2, 6.A.3, 6.A.4
	14.3	6.A.3
	14.4	6.A.1, 6.A.2, 6.A.3, 6.A.4
	14.5	6.B.1, 6.B.2
Chapter 15 Acids and Bases pp. 149–159	15.1	6.C.1
	15.2	6.C.1, 6.C.2
	15.3	6.C.1, 6.C.2
	15.4	6.C.1, 6.C.2
	15.5	6.C.1, 6.C.2
	15.6	6.C.1, 6.C.2
	15.7	6.A.4, 6.C.1
	15.8	6.A.4, 6.C.1, 6.C.2
	15.9	6.C.1
	15.10	6.A.4, 6.C.1, 6.C.2
	15.11	6.C.1, 6.C.2
	15.12	Prerequisite Knowledge

Chapter	Section	AP Correlation
Chapter 16 Acid–Base Equilibria and Solubility Equilibria pp. 160–174	16.1	6.C.1, 6.C.3
	16.2	6.B.1, 6.B.2, 6.B.4
	16.3	6.C.1, 6.C.2
	16.4	1.E.2, 3.A.2, 3.B.2, 6.C.1
	16.5	3.B.2, 6.C.1
	16.6	6.C.3
	16.7	6.B.1, 6.C.3
	16.8	6.B.1, 6.B.2, 6.C.3
	16.9	6.B.1
	16.10	6.B.1, 6.B.2
	16.11	6.B.1, 6.B.2, 6.C.3
Chapter 17 Entropy, Free Energy, and Equilibrium pp. 175–185	17.1	5.B.2
	17.2	5.E.2, 5.E.5
	17.3	5.E.1
	17.4	5.E.1, 5.E.2
	17.5	5.E.2, 5.E.3, 5.E.4
	17.6	6.D.1
	17.7	5.E.4
Chapter 18 Electrochemistry pp. 186–196	18.1	3.B.3, 3.C.3
	18.2	3.C.3
	18.3	3.C.3
	18.4	3.C.3
	18.5	Extending Knowledge
	18.6	3.C.3
	18.7	3.B.3
	18.8	3.C.3, 5.E.4
Chapter 19 Nuclear Chemistry pp. 197–201	19.1	Extending Knowledge
	19.2	Extending Knowledge
	19.3	4.A.3
	19.4	Extending Knowledge
	19.5	Extending Knowledge
	19.6	Extending Knowledge
	19.7	Extending Knowledge
	19.8	Extending Knowledge

Chapter	Section	AP Correlation
Chapter 20 Chemistry in the Atmosphere pp. 900–929		Not covered by the AP Curriculum and not included in the AP Focus Review Guide
Chapter 21 Metallurgy and the Chemistry of Metals pp. 202–206	21.1	Extending Knowledge
	21.2	Extending Knowledge
	21.3	2.D.2, 2.D.3
	21.4	1.C.1
	21.5	1.C.1
	21.6	1.C.1, 3.C.3
	21.7	3.C.3
Chapter 22 Nonmetallic Elements and Their Compounds pp. 207–209	22.1	1.C.1
	22.2	Prerequisite Knowledge
	22.3	Prerequisite Knowledge
	22.4	Prerequisite Knowledge
	22.5	Prerequisite Knowledge
	22.6	1.C.1
Chapter 23 Transition Metals Chemistry and Coordination Compounds pp. 210–214	23.1	1.C.1
	23.2	Extending Knowledge
	23.3	Extending Knowledge
	23.4	Extending Knowledge
	23.5	Extending Knowledge
	23.6	Extending Knowledge
	23.7	Extending Knowledge
Chapter 24 Organic Chemistry pp. 1025–1057		Not covered by the AP Curriculum and not included in the AP Focus Review Guide
Chapter 25 Synthetic and Natural Organic Polymers pp. 215–219	25.1	Extending Knowledge
	25.2	Extending Knowledge
	25.3	2.B.2, 5.D.3
	25.4	2.B.2, 5.D.3

Using Your AP Focus Review Guide

This review guide was developed with the AP student in mind. The activities within each chapter will help you to focus on and review the key content in the chapter as it relates to the AP Chemistry.

The **AP A Look Ahead** boxes review the Big Ideas covered in your textbook.

These correlations pinpoint which parts of the AP Curriculum are reviewed in each section.

Review It activities will help you to recall content that you should have mastered in earlier study.

Different visual organizers help you to analyze and summarize information and remember content.

Use It activities allow you to apply what you've learned in the chapter.

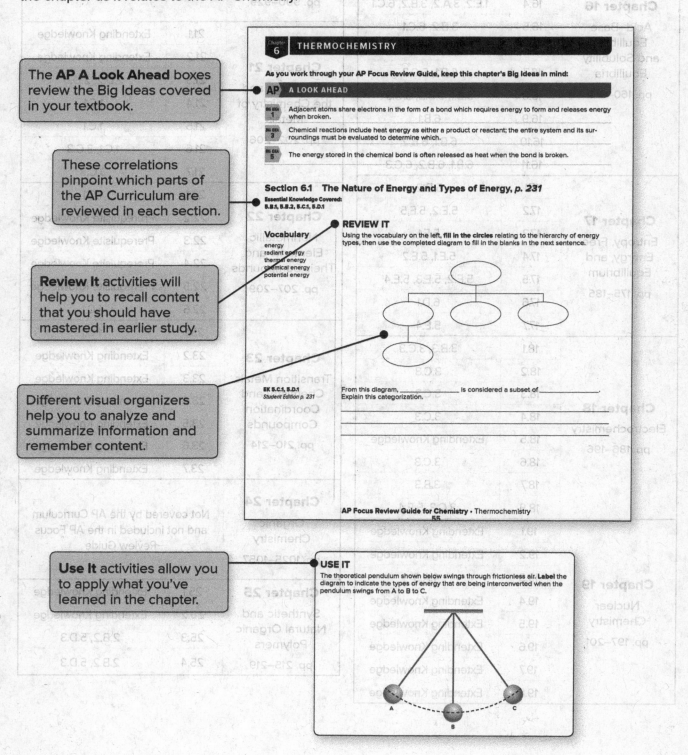

In the **Summarize It** section you will synthesize what you have learned by focusing in on the key content in the chapter.

You can use this column to take notes and keep track of content you may need to review in preparation for the exam.

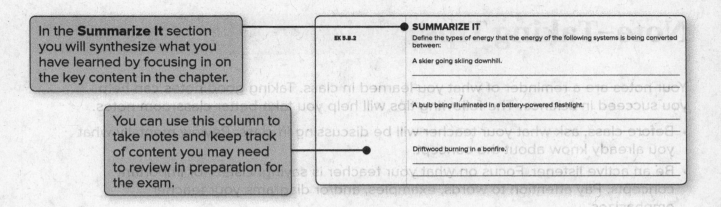

SUMMARIZE IT

EK 5.B.2

Define the types of energy that the energy of the following systems is being converted between:

A skier going skiing downhill.

A bulb being illuminated in a battery-powered flashlight.

Driftwood burning in a bonfire.

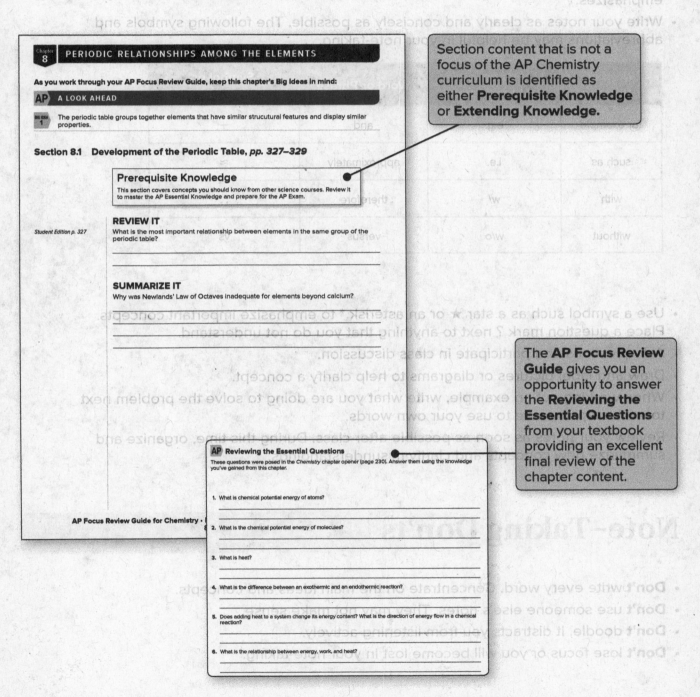

Section content that is not a focus of the AP Chemistry curriculum is identified as either **Prerequisite Knowledge** or **Extending Knowledge.**

The **AP Focus Review Guide** gives you an opportunity to answer the **Reviewing the Essential Questions** from your textbook providing an excellent final review of the chapter content.

Chapter 8 — PERIODIC RELATIONSHIPS AMONG THE ELEMENTS

As you work through your AP Focus Review Guide, keep this chapter's Big Ideas in mind:

AP A LOOK AHEAD

BIG IDEA 1 — The periodic table groups together elements that have similar structural features and display similar properties.

Section 8.1   Development of the Periodic Table, pp. 327–329

Prerequisite Knowledge

This section covers concepts you should know from other science courses. Review it to master the AP Essential Knowledge and prepare for the AP Exam.

REVIEW IT

Student Edition p. 327

What is the most important relationship between elements in the same group of the periodic table?

SUMMARIZE IT

Why was Newlands' Law of Octaves inadequate for elements beyond calcium?

AP Focus Review Guide for Chemistry •

AP Reviewing the Essential Questions

These questions were posed in the *Chemistry* chapter opener (page 230). Answer them using the knowledge you've gained from this chapter.

1. What is chemical potential energy of atoms?

2. What is the chemical potential energy of molecules?

3. What is heat?

4. What is the difference between an exothermic and an endothermic reaction?

5. Does adding heat to a system change its energy content? What is the direction of energy flow in a chemical reaction?

6. What is the relationship between energy, work, and heat?

Note–Taking Tips

Your notes are a reminder of what you learned in class. Taking good notes can help you succeed in science. The following tips will help you take better classroom notes.

- Before class, ask what your teacher will be discussing in class. Review mentally what you already know about the concept.
- Be an active listener. Focus on what your teacher is saying. Listen for important concepts. Pay attention to words, examples, and/or diagrams your teacher emphasizes.
- Write your notes as clearly and concisely as possible. The following symbols and abbreviations may be helpful in your note-taking.

Word or Phrase	Symbol or Abbreviation	Word or Phrase	Symbol or Abbreviation
for example	e.g.	and	+
such as	i.e.	approximately	≈
with	w/	therefore	∴
without	w/o	versus	vs

- Use a symbol such as a star ★ or an asterisk * to emphasize important concepts. Place a question mark ? next to anything that you do not understand.
- Ask questions and participate in class discussion.
- Draw and label pictures or diagrams to help clarify a concept.
- When working out an example, write what you are doing to solve the problem next to each step. Be sure to use your own words.
- Review your notes as soon as possible after class. During this time, organize and summarize new concepts and clarify misunderstandings.

Note–Taking Don'ts

- **Don't** write every word. Concentrate on the main ideas and concepts.
- **Don't** use someone else's notes. They may not make sense.
- **Don't** doodle. It distracts you from listening actively.
- **Don't** lose focus or you will become lost in your note-taking.

As you work through your AP Focus Review Guide, keep this chapter's Big Ideas in mind:

AP   A LOOK AHEAD

 Pure substances will be classified as elements or as compounds.

 The physical and chemical properties of matter are connected to its atomic and molecular structure.

Section 1.1   Chemistry: A Study for the Twenty-First Century, *p. 2*

Prerequisite Knowledge

This section covers concepts you should know from other science courses. Review it to master the AP Essential Knowledge and prepare for the AP Exam.

USE IT

Student Edition p. 2

Give an example of how chemistry relates to or is used in these other scientific disciplines you might study in high school:

Environmental Science:

_____

Physics:

_____

Biology:

_____

Section 1.2   The Study of Chemistry, *pp. 2–3*

Prerequisite Knowledge

This section covers concepts you should know from other science courses. Review it to master the AP Essential Knowledge and prepare for the AP Exam.

USE IT

Student Edition p. 3

For the following observations, indicate if they are macroscopic (MA) or microscopic (MI).

A piece of paper burns when lit with a match. _____

Copper will oxidize over time. _____

Iron reacts with oxygen in the presence of water. _____

Prerequisite Knowledge

This section covers concepts you should know from other science courses. Review it to master the AP Essential Knowledge and prepare for the AP Exam.

USE IT

Student Edition p. 4

Label the following observations as either qualitative or quantitative.

1. The metal bar was 15 cm long. _____

2. The solid material was hard to the touch. _____

3. The temperature of the metal bar was 21°C. _____

4. The surface of the metal was shiny. _____

5. The metal object was grayish-silver in color. _____

6. The metal bar had a mass of 224-g. _____

7. The metal bar heated quickly. _____

8. It took 17 minutes for the bar to cool from 35°C to 21°C. _____

Section 1.4   Classifications of Matter, *pp. 6–8*

**Essential Knowledge
Covered: 1.A.1, 2.A.3**

REVIEW IT

Word Bank

compound
element
heterogeneous mixture
homogeneous mixture
matter
mixture
substance

EK 2.A.3
Student Edition pp. 6–7

Using the word bank on the left, **fill in the blanks.**

Any material that occupies space and has a mass is referred to as _____.

A form of this that has a definite composition and distinct properties is a(n)

_____. When two or more of these materials combine, yet retain their

properties, the combination is called a(n) _____. This combination is

referred to as a(n) _____ if the composition is the same through-

out and a(n) _____ if the composition is not uniform. Both of these

can be separated into their original components by various methods. However, if the

material cannot be separated into simpler substances by chemical means, the material

is referred to as a(n) _____. Conversely, a(n) _____ can be sepa-

rated into simpler substances by chemical means.

Section 1.4 Classifications of Matter (continued)

USE IT

EK 2.A.3
Student Edition pp. 6–7

Place the terms *matter, mixtures, pure substances, compounds, heterogeneous mixtures, elements* and *homogeneous mixtures* in the following flow chart, and describe separation methods between the boxes connected with the arrows:

SUMMARIZE IT

EK 1.A.1
Student Edition p. 8

Identify each as a compound or an element.

water _____

iron _____

Solid carbon in the form of a diamond _____

$CaCl_2$ _____

Section 1.5   The Three States of Matter, *pp. 9–10*

Essential Knowledge Covered:
1.A.1, 2.A.1, 2.A.2

REVIEW IT

EK 2.A.1, 2.A.2
Student Edition p. 9

Beside each statement, place the term solid, liquid or gas that correctly applies to the statement.

Molecules are tightly held together with little room for motion. _____

Molecules are close together, but can move past one another with little restriction. _____

Molecules are far apart and free to move in all directions. _____

EK 1.A.1
Student Edition p. 9

Complete the following concept chart by filling in the missing term or definition.

Freezing point	
	The point where a liquid becomes a gas.
Melting Point	

EK 1.A.1
Student Edition p. 9

Agar is a substance that melts at 85°C and solidifies from 31°C to 40°C. What does this indicate about freezing and melting points?

_____

EK 1.A.1
Student Edition p. 9

What is true of the composition of a compound as it changes state?

_____

_____

USE IT

EK 2.A.1, 2.A.2
Student Edition p. 9

Draw three diagrams showing what the atomic arrangement of a solid, liquid, and gaseous matter contained inside a beaker.

SUMMARIZE IT

EK 2.A.1, 2.A.2
Student Edition pp. 9–10

Use the flowchart below to summarize the properties of the three states of matter. In the column to the left, indicate the macroscopic properties of each state of matter. In the column to the right, include properties associated with the particles of the substance when in that state of matter on a microscopic level.

Solid

fixed _____ and _____

particles are _____

Liquid

fixed _____, flexible

particles are _____
_____

Gas

takes on the _____
_____

particles are _____
_____

Section 1.6 Physical and Chemical Properties of Matter, *pp. 10–11*

USE IT

Student Edition pp. 10–11

Below, you will find a diagram of a compound made up of two elements, each represented using spheres of different shades. Below each, create a diagram to show each of the indicated changes:

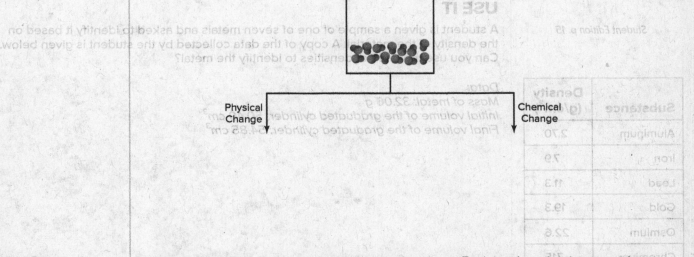

Physical Change

Chemical Change

Student Edition p. 11

When a candle burns, the wax can either melt or burn. Explain why scientists consider this an *ambiguous change*.

_____

_____

_____

Section 1.7   Measurement, *pp. 11–18*

Section 1.8   Handling Numbers, *pp. 18–23*

Section 1.9   Dimensional Analysis in Solving Problems, *pp. 22–27*

Section 1.10   Real-World Problem Solving: Information, Assumption and Simplifications, *pp. 27–28*

Extending Knowledge

This section takes the AP Essential Knowledge you have learned further, and may provide illustrative examples useful for the AP Exam.

USE IT

Student Edition p. 15

A student is given a sample of one of seven metals and asked to identify it based on the density of the material. A copy of the data collected by the student is given below. Can you use the table of densities to identify the metal?

Data:
Mass of metal: 32.06 g
Initial volume of the graduated cylinder: 50.40 cm³
Final volume of the graduated cylinder: 54.85 cm³

Substance	Density (g/cm³)
Aluminum	2.70
Iron	7.9
Lead	11.3
Gold	19.3
Osmium	22.6
Chromium	7.15
Tin	7.27

USE IT

Student Edition p. 23

A dart player throws three darts and all three dots land in the bull's eye. However, the dart player was aiming for the triple 20 with all three darts. Comment on the precision and accuracy of the player.

Student Edition pp. 18–19

The table below shows the volumetric tolerances for a variety of chemical glassware.

Tolerances for Volumetric Ware

Maximum Allowed Error, mL			
Volume, mL	Volumetric Flask	Volumetric Pipet	Burette
5	–	0.01	–
10	–	0.02	–
25	0.03	0.03	0.03
50	0.05	0.05	0.05
100	0.08	0.08	0.10
250	0.10	–	–
500	0.15	–	–
1000	0.30	–	–

*Data from J. S. Fritz and G. H. Schenk. *Quantitative Analytical Chemistry,* 3rd edition. Allyn & Bacon, Boston, 1974. p. 560.

If you filled a 25 mL volumetric pipet exactly to the line, what range of values would the volume exist between?

_____

Which piece of equipment would best be used to measure exactly 25 mL of solution: a volumetric flask or a pipette? Explain your answer.

_____

Student Edition pp. 20–22

Calculate the final answer to the correct number of significant figures.

86.25 g + 14.3 g = _____

16.4 g/cm^3 × 1.35 cm^3 = _____

3 bars of gold × 25.3 kg per bar = _____

USE IT

Student Edition p. 24

You plan on travelling to Europe this summer and wish to convert some of your US dollars ($) to the Euro (€).

You are told that 1 US dollar is the equivalent to 0.73 Euro.

How many Euro would you get out of $450?

How much would you need in dollars to be able to convert have 1000€?

Student Edition p. 24

If 8 gold coins weigh 43.5 g, what would 31 gold coins weigh?

USE IT

Fill in the sequence of unit conversions needed to change from fl oz to m³.

Given that there are 0.033814 fl oz in 1 mL, use your sequence from question 1 to convert 828 fl oz to m³.

Explain how you arrived at the number of significant figures in this answer.

_____

_____

_____

USE IT

A pile of gravel is in the approximate shape of an inverted cone, with a diameter of 18 m and a height of 34 m. Gravel has a density of 1255 kg/m³.

Given the formula for the volume of a cone $V = \frac{1}{3}\pi r^2 h$, what assumptions can you make in your calculation for the volume of the pile of gravel?

_____

_____

_____

Use this approximation to find the volume of the gravel pile.

Using your answer to the above question and the density of the gravel, determine the approximate mass of the gravel pile.

What assumptions were made in this calculation?

_____

_____

What effect do you believe your assumptions had on your final answer? Explain.

_____

_____

AP Reviewing the Essential Questions

These questions were posed in the *Chemistry* chapter opener (page 1). Answer them using the knowledge you've gained from this chapter.

1. What is an element? Can atoms combine together?

_____

_____

2. What is a mixture?

_____

_____

3. What are solids and liquids?

_____

_____

4. What is a gas?

_____

_____

Chapter 2

ATOMS, MOLECULES, AND IONS

As you work through your AP Focus Review Guide, keep this chapter's Big Ideas in mind:

AP   A LOOK AHEAD

BIG IDEA 1 Atomic theory, which has been built over time by many scientists, describes the structure of the atom.

BIG IDEA 2 The fundamental structure of the atom defines how atoms come together to form higher order structures.

Section 2.1   The Atomic Theory, *pp. 39–40*

Essential Knowledge Covered:
1.A.1, 1.D.1

REVIEW IT

Using the word bank on the left, **fill in the blanks:**

Word Bank

atoms
chemical
compounds
destruction
different
elements
fraction
identical
integer
mass
more
ratio
reaction
rearrangement

The four essential hypotheses about the nature of matter:

1. _____ are composed of extremely small particles called _____.

2. All atoms of a given element are _____, having the same size, _____, and _____ properties. The atoms of one element are _____ from the atoms of all other elements.

3. _____ are composed of atoms of _____ than one element. In any compound, the _____ of the numbers of atoms of any two of the elements present is either an _____ or a simple _____.

4. A chemical _____ involves only the separation, combination, or _____ of atoms; it does not result in their creation or _____.

EK 1.A.1, 1.D.1
Student Edition p. 40

SUMMARIZE IT

The Greek Philosopher Democritus named the small, indivisible particles that he thought all things were made of 'atomos,' which means 'indivisible' or 'uncuttable.' Would he still choose that name if he were alive today and knew what you know? Why or why not?

_____

_____

_____

Section 2.2   The Structure of the Atom, *pp. 40–46*

Essential Knowledge Covered:
1.B.1, 1.D.1

REVIEW IT

Vocabulary

electron
element
experimentation
model
neutron
nucleus
orbital
proton
shell

Using the vocabulary on the left, **fill in the blanks:**

An atom is the basic unit of an _____ that can enter into a chemical combina-

tion. The currently held _____ of the atom is the result of many years of

_____ by scientists all over the world. The atom is composed of three

fundamental parts: a negatively charged _____ which exists outside the

_____ in a _____ or _____, a _____ located

inside the nucleus, and an uncharged particle called a _____ which is also

located in the nucleus.

USE IT

EK 1.B.1
Student Edition p. 46

Complete the following table with respect to the fundamental subatomic particles:

Particle	Mass (g)	Charge (C)	Unit of charge
Proton			
Electron			
Neutron			

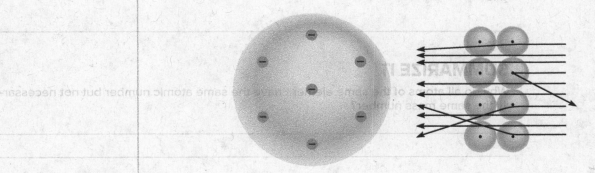

EK 1.D.1
Student Edition pp. 43–44

Label the regions of positive and negative charge in both Thomson's 'Plum Pudding' model and Rutherford's Gold Foil Experiment. Now, imagine that both Rutherford and Thomson exist in 2014 and that Rutherford wants to inform Thomson of his discovery using social media. In 140 characters or less, what would Rutherford say?

_____

_____

_____

Section 2.2   The Structure of the Atom (continued)

SUMMARIZE IT

EK 1.D.1

Why do scientists refer to Atomic Theory and not Atomic Law?

_____

_____

Section 2.3   Atomic Number, Mass Number, and Isotopes, *pp. 46–47*
Essential Knowledge Covered:
1.D.1, 1.D.2

REVIEW IT

EK 1.D.2
Student Edition p. 46

On the periodic table, the two-letter atomic abbreviation is typically shown with one two numbers beside it. Define what each number means and how they are calculated.

_____

_____

USE IT

EK 1.D.1
Student Edition p. 46

Most elements have more than one isotope, which is an atom that has the same atomic number but a different mass number. How many protons and neutrons are in the following isotope of He (helium four)?

^4_2He

_____

SUMMARIZE IT

EK 1.D.1

Why do all atoms of the same element have the same atomic number but not necessarily the same mass number?

_____

_____

_____

Section 2.4   The Periodic Table, *pp. 48–49*

Essential Knowledge Covered:
1.C.1

REVIEW IT

EK 1.C.1
Student Edition p. 50

Several groups have characteristic names. Give an example of two elements in each group/family:

Alkali Metals: _____

Alkaline Earth Metals: _____

Halogens: _____

Noble Gases: _____

USE IT

EK 1.C.1
Student Edition p. 48

Using the periodic table on page 48 of your textbook, determine what category of elements comprises the majority of the periodic table.

_____

_____

EK 1.C.1
Student Edition p. 48

How do metals differ from nonmetals?

_____

_____

SUMMARIZE IT

EK 1.C.1

You are building a device that uses, in its construction, a significant amount of magnesium. You notice that your supply of Mg has run out, but you have aluminum as well as beryllium available. Explain which element would be the best element to use as a substitute, given that you do not want to change the device's physical properties?

_____

_____

_____

Section 2.5   Molecules and Ions, *pp. 50–51*

Essential Knowledge Covered:
1.A.1

USE IT

EK 1.A.1
Student Edition p. 51

Referring to the Periodic Table on page 51 of your textbook, how many hydroxide ions would need to combine with a single cation of the given element to form a neutral compound? List the number necessary for each monatomic ion possible for each element, or "No Reaction" if the listed ion would not combine with a hydroxide ion.

Fe:

Na:

Mn:

Sn:

N:

EK 1.A.1
Student Edition p. 51

In the above example, what is common between all the ions that will form neutral compounds with a hydroxide ion?

_____

EK 1.A.1
Student Edition p. 51

A subscripted number immediately following an atomic symbol indicates the number of that species present in a molecule. What is the ratio of hydrogen atoms to oxygen atoms in the following compounds:

H_3PO_4:

H_2O:

H_2O_2:

SUMMARIZE IT

EK 1.A.1

Turn to the periodic table shown on page 51 of your textbook. It appears that only elements coded in green tend to form cations, whereas elements in blue tend to form anions. What general trend about ionic bonds can you conclude from this?

_____

_____

EK 1.A.1

Using the periodic table on page 51, do you see any trends relating to the preferred cationic or anionic charge and principle group number? The principle group numbers are followed by the letter A.

_____

_____

Section 2.6   Chemical Formulas, *pp. 52–55*

Essential Knowledge Covered:
1.A.1, 2.C.2

REVIEW IT

EK 1.A.1
Student Edition p. 53

How do the empirical formulas for water (H_2O) and hydrogen peroxide (H_2O_2) compare? How do they differ?

_____

_____

_____

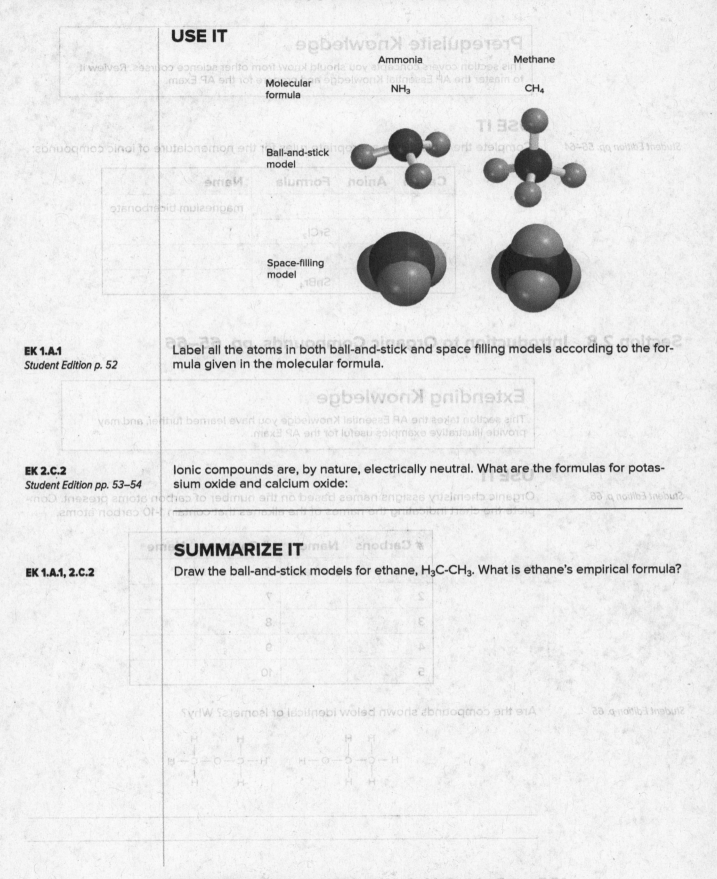

USE IT

Ammonia

Methane

Molecular formula

NH_3

CH_4

Ball-and-stick model

Space-filling model

EK 1.A.1
Student Edition p. 52

Label all the atoms in both ball-and-stick and space filling models according to the formula given in the molecular formula.

EK 2.C.2
Student Edition pp. 53–54

Ionic compounds are, by nature, electrically neutral. What are the formulas for potassium oxide and calcium oxide:

SUMMARIZE IT

EK 1.A.1, 2.C.2

Draw the ball-and-stick models for ethane, $H_3C\text{-}CH_3$. What is ethane's empirical formula?

Section 2.7   Naming Compounds, *pp. 56–65*

USE IT

Student Edition pp. 56–64

Complete the table, using appropriate rules for the nomenclature of ionic compounds:

Cation	Anion	Formula	Name
			magnesium bicarbonate
		$SrCl_2$	
$Fe3^+$	NO_2^-		
		$SnBr_4$	

Section 2.8   Introduction to Organic Compounds, *pp. 65–66*

USE IT

Student Edition p. 66

Organic chemistry assigns names based on the number of carbon atoms present. Complete the chart indicating the names of the alkanes that contain 1-10 carbon atoms.

# Carbons	Name	# Carbons	Name
1	methane	6	
2		7	
3		8	
4		9	
5		10	

Student Edition p. 65

Are the compounds shown below identical or isomers? Why?

_____

_____

These questions were posed in the *Chemistry* chapter opener (page 38). Answer them using the knowledge you've gained from this chapter.

1. How does an atom differ from a molecule?

   _____

2. What developments have changed our understanding of atoms and molecules since Dalton's proposed atomic theory?

   _____
   _____
   _____

3. What particles comprise an atom?

   _____
   _____
   _____

4. How do elements differ at the atomic level? Are all atoms of a given element identical?

   _____
   _____
   _____

5. What is the basic structure of the periodic table?

   _____
   _____

6. Can atoms form ions? How do ionic compounds maintain electrical neutrality?

   _____
   _____
   _____

As you work through your AP Focus Study Guide, keep this chapter's Big Ideas in mind:

AP A LOOK AHEAD

BIG IDEA 1 An understanding of atomic mass, molar mass and percent composition is important for performing calculations involving chemical reactions.

BIG IDEA 3 Determining the amounts of products that can form when given amounts of reactants are used allows for the calculation of reaction yields.

Section 3.1   Atomic Mass, *pp. 76–77*

Essential Knowledge Covered:
1.A.3

REVIEW IT

Student Edition p. 76

Using the word bank on the left, **fill in the blanks**. Words may be used more than once.

Word Bank

atomic mass
atomic mass unit
amu
average atomic mass
carbon
isotope
neutron
proton

Scientists have established a relative scale for atoms that uses _____-12 as the standard. This is the _____ of carbon that contains six _____ and six _____. One _____ is defined as the mass of one-twelfth the mass of one atom of _____-12. This unit is abbreviated as the _____. The _____ of all other elements is based on this value.

The weighted average of all of the naturally occurring _____ of an element is the _____ for the element. The reason why some elements have more than isotope is due to a difference in the number of _____ for the atoms, while the number of _____ remain the same for all isotopes of an element.

USE IT

EK 1.A.3
Student Edition p. 77

An element has been found that has three naturally occurring isotopes as given in the chart, the most abundant of which is an atom with an equal number of protons and neutrons.

Isotope	Mass (amu)	Abundance (%)
A	22.00	82.4
B	23.01	15.2
C	25.02	2.4

EK 1.A.3
Student Edition p. 77

Use this information in the table on p. 18 to determine the average atomic mass for the element.

EK 1.A.3
Student Edition p. 76

Why is the average atomic mass closer to the atomic mass of isotope A than it is to isotope B or isotope C?

_____

_____

EK 1.A.3

| 5 |
| **B** |
| 10.81 |

SUMMARIZE IT

The element box to the left is that of boron. It has two naturally occurring isotopes: boron-10 and boron-11.

Fill in the table for each isotope of boron:

Isotope	Number of protons	Number of neutrons
Boron-10		
Boron-11		

Boron-10 has an abundance of 19.9%. Verify the value of the average atomic mass.

Explain any difference between your calculated value and the value in the element box.

_____

_____

Section 3.2 Avogadro's Number and the Molar Mass of an Element, *pp. 77–81*

Essential Knowledge Covered:
1.A.3

REVIEW IT

EK 1.A.3
Student Edition pp. 77–78

Fill in the blanks below. Do not use a calculator on this problem.

12.00 g of carbon-12 is 1 mol, so 36.00 g of the material would be _____ mol and

1.20 g would be _____ mol.

EK 1.A.3
Student Edition pp. 79–80

USE IT

Find the periodic table in the front of your textbook. How many moles of atoms are in a 2.50 g sample of magnesium?

SUMMARIZE IT

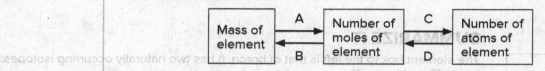

EK 1.A.3

In the flow chart above, there are four arrows labeled A, B, C, and D. Each of these arrows represents a mathematical operation that must be done to move from value to value. What mathematical operation is needed for each numbered step?

A: _____

B: _____

C: _____

D: _____

EK 1.A.3

Use the flow chart above to determine the number of atoms in a 334.21 g sample of iron.

5
Fe
55.845

Section 3.3   Molecular Mass, *pp. 81–83*

Essential Knowledge Covered:
1.A.3

REVIEW IT

EK 1.A.3
Student Edition p. 81

Determine the number of each type of atom in one molecule of diphosphorus pentoxide.

_____

USE IT

EK 1.A.3
Student Edition p. 81

For each compound, determine the molecular masses (in amu).

Acetylsalicylic acid, $C_9H_8O_4$, the key ingredient in many pain relievers

Section 3.3   Molecular Mass (continued)

Aluminum chloride, $AlCl_3$, added to many antiperspirants

SUMMARIZE IT

EK 1.A.3

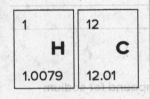

1	12
H	**C**
1.0079	12.01

The structure of propane is given to the left. You are given a sample of 255.30 g of propane and are asked to determine the number of moles this sample contains.

Determine the number of moles in the 255.30 g sample.

Section 3.4   The Mass Spectrometer, *pp. 83–85*

Essential Knowledge Covered:
1.D.2

SUMMARIZE IT

An element exists with three naturally occurring isotopes:

Isotope	Mass (amu)	Abundance (%)
A	82.00	78.2
B	85.05	17.8
C	87.10	4.0

EK 1.D.2

Sketch a mass spectrum for this element.

Explain how you arrived at your mass spectrum

_____

_____

_____

Section 3.5 Percent Composition of Compounds, *pp. 85–88*

Essential Knowledge Covered:
1.A.1, 1.A.2, 1.A.3

Student Edition pp. 85–88
Student Edition p. 86

Word Bank

empirical formula
mass percent
moles of each element
mole ratios of elements

REVIEW IT

Fill in the blanks for the flow chart that can be used to find the empirical formula when given percent composition by mass for each element in a compound.

Identify what must be done mathematically to move from step to step:

_____

_____

_____

USE IT

EK 1.A.1, 1.A.2, 1.A.3
Student Edition p. 89

Determine the percent composition for each element in the compound for sodium carbonate, Na_2CO_3.

EK 1.A.1, 1.A.2, 1.A.3
Student Edition p. 89

A molecule is made up of 50.05% sulfur and 49.95% oxygen. Use the flow chart from the REVIEW IT section to determine the empirical formula of the compound.

SUMMARIZE IT

EK 1.A.1, 1.A.2, 1.A.3

Two large deposits of aluminum were found in two areas where a mining company was looking to open a new aluminum mine. Site A has aluminum in the form of Al_2O_3 while at Site B, the aluminum is on the form of $KAl(SO_4)_2$.

What considerations would the mining company use to make a decision as to where to open a mining operation?

Determine what ratio of mass deposits would be needed for both sites to be considered as sites to open a mining operation.

_____

_____

Section 3.6 Experimental Determination of Empirical Formulas, *pp. 88–90*

Essential Knowledge Covered:
1.A.2, 1.A.3

REVIEW IT

EK 1.A.2
Student Edition pp. 88–89

Identify which are possible molecular formulae for the empirical formula CH_2.

C_2H_4 C_2H_6 C_5H_{10} C_3H_8 C_6H_6 CH_4 C_3H_6

SUMMARIZE IT

EK 1.A.2, 1.A.3

A compound containing only carbon and hydrogen was burned in an apparatus that can be used to detect the amount of carbon dioxide and water produced in the process. When 15.00 g of the material is burned, 50.78 g of carbon dioxide and 10.39 g of water are produced. The molar mass of this compound is found to be approximately 26 g.

Create flow charts to illustrate the process needed to find the molecular formula:

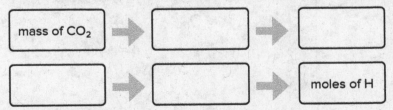

mass of CO_2 ➡️ [   ] ➡️ [   ]

[   ] ➡️ [   ] ➡️ moles of H

Use this flow chart to find the molecular formula for the compound.

Section 3.7   Chemical Reactions and Chemical Equations, *pp. 90–95*

Essential Knowledge Covered:
1.A.1, 1.E.1, 1.E.2, 3.A.1, 3.C.1

USE IT

EK 3.A.1, 1.E.1, 1.E.2
Student Edition p. 93

Balance the combustion reaction of propane, C_3H_8. In this reaction, propane combines with oxygen gas to produce carbon dioxide and water.

___ C_3H_8 + ____O_2 → ____CO_2 + _____H_2O

Fill in the chart for a balance sheet of atoms for the reaction above:

Element	Number on reactant side	Number on product side
Carbon		
Hydrogen		
Oxygen		

EK 3.A.1, 1.E.1, 1.E.2
Student Edition p. 93

Another student balanced the reaction as

$3\ C_3H_8 + 15\ O_2 \rightarrow 9\ CO_2 + 12\ H_2O$

What would you say to this student about their answer? Be specific and correct any issues that need correcting in what they did.

_____

SUMMARIZE IT

EK 3.A.1

Balance each chemical reaction given here, and show the balance sheet for the number of each type of atom once each reaction has been balanced.

$Fe + Cl_2 \rightarrow FeCl_3$

Element	Number on reactant side	Number on product side
Iron		
Chlorine		

$Na + H_2O \rightarrow NaOH + H_2$

Element	Number on reactant side	Number on product side
Sodium		
Hydrogen		
Oxygen		

$C_6H_6 + O_2 \rightarrow CO_2 + H_2O$

Element	Number on reactant side	Number on product side
Carbon		
Hydrogen		
Oxygen		

Section 3.8   Amounts of Reactants and Products, *pp. 95–99*

Essential Knowledge Covered:
1.A.2, 1.A.3, 1.E.1, 1.E.2, 3.A.1, 3.A.2

REVIEW IT

Using the word bank on the left, **fill in the blanks:**

Word Bank

coefficient
stoichiometry
mole method
moles

The quantitative study of reactants and products in a chemical reaction is the study of

_____. All calculations that use the balanced chemical reaction require that

the _____ in the reaction be read as the number of _____ of each

substance. For this reason, it is referred to as the _____.

EK 1.E.1, 1.E.2
Student Edition p. 95

Write the reaction $2H_2O(l) \rightarrow 2H_2(g) + O_2(g)$ in words.

_____

_____

EK 1.E.1, 1.E.2
Student Edition p. 95

Balance the reaction $P(s) + O_2(g) \rightarrow P_2O_5(s)$

Once balanced, draw a representation of the balanced chemical reaction using black spheres for the P atoms and white spheres for the O atoms.

EK 1.A.2, 1.A.3
Student Edition p. 97

Word Bank

Mass of X
Moles of X
Moles of Y
Mass of Y

USE IT

Use the work bank to fill in the graphic organizer for calculating the mass of product Y when you are given the mass of reactant X.

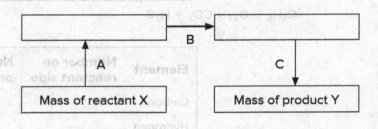

EK1.A.2, 1.A.3
Student Edition p. 97

Explain what mathematically needs to be done at steps A, B, and C to move through the organizer.

_____

_____

SUMMARIZE IT

EK 1.A.3, 3.A.1, 3.A.2
Student Edition p. 97

Water decomposes into hydrogen gas and oxygen gas when electricity passes through the solution. If 24.50 g of water is fully decomposed, what mass of oxygen will be collected?

Section 3.9   Limiting Reagents, *pp. 99–102*

Essential Knowledge Covered:
1.A.3, 3.A.1, 3.A.2

EK 3.A.1, 3.A.2
Student Edition
pp. 100–101

REVIEW IT

In the process of assembling tricycles for small children, a worker has 16 frames and 60 tires with which to assemble frame and tire components.

Write a balanced equation for this process.

In this situation, which would be considered the limiting reagent and which would be the excess reagent?

_____

_____

_____

EK 1.A.3, 3.A.1, 3.A.2
Student Edition
pp. 100–101

USE IT

In the reaction of $Fe(s) + O_2(g) \rightarrow Fe_2O_3(s)$,

What is the exact stoichiometric ratio of iron to oxygen in the reaction (be sure to balance the reaction before answering)?

_____

_____

If 2 moles of iron and 2 moles of oxygen gas were to be used, which would be the limiting reagent? Explain your answer.

_____

_____

_____

EK 1.A.3, 3.A.1, 3.A.2

28
Ni
58.69

Molar mass (g/mol):
Lactic acid 90.08

SUMMARIZE IT

Determine the mass of hydrogen gas that will form when 22.53 g of solid nickel reacts with 65.16 g of lactic acid in the reaction:

$Ni(s) + HC_2H_5O_3(aq) \rightarrow Ni(C_2H_5O_3)_2 \, (aq) + H_2(g)$

Show all work and be sure to balance the chemical reaction before you begin the calculation.

Essential Knowledge Covered:
1.A.1, 1.A.3, 3.A.1, 3.A.2

EK 1.A.1, 1.A.3
Student Edition p. 103

Word Bank

actual yield
theoretical yield
percent yield

REVIEW IT

Using the word bank on the left, **match the expressions to the terms** to review the Learning Outcomes of this section:

The proportion used to determine the efficiency of a chemical reaction:

The amount of product that will form once the limiting reagent is completely converted to product:

The amount of product that is found to experimentally form when a chemical reaction is allowed to occur:

EK 1.A.3, 3.A.1, 3.A.2
Student Edition p. 104

Molar/Atomic masses (g/mol):
SiO_2: 60.08
C: 12.01
SiC: 40.11
CO_2: 44.01

SUMMARIZE IT

Silicon carbide, SiC(s) can be manufactured by reacting silicon dioxide with solid carbon at high temperatures. The second product in this reaction is carbon dioxide gas. The reaction can be written as

$SiO_2(s) + 3\ C(s) \rightarrow SiC(s) + 2\ CO_2(g)$

If 20.00 g of silicon dioxide is reacted with 14.02 g of carbon, 12.05 g of silicon carbide forms.

Determine which reactant is the limiting reagent in this process.

How many moles of silicon carbide will form in this process?

What is the theoretical yield in this situation?

Determine the reaction yield.

EK 3.A.1, 3.A.2   A classmate said that they read an article stating that the percent yield of a new chemical reaction is more than 100%. What would you say to this classmate in response to this claim?

_____

_____

_____

_____

AP Reviewing the Essential Questions

These questions were posed in the *Chemistry* chapter opener (page 75). Answer them using the knowledge you've gained from this chapter.

1. How do we determine atomic mass? What is a mole? How do we determine molar mass?

_____

_____

_____

2. How does mass spectroscopy demonstrate the existence of isotopes?

_____

_____

_____

3. What is the percent elemental composition of a substance? How do we experimentally determine composition?

_____

_____

_____

4. What is conservation of mass?

_____

_____

_____

5. How do we correctly represent a chemical change?

_____

_____

_____

6. How do we write a properly balanced chemical reaction?

_____

_____

_____

7. What evidence is used to indicate a chemical change has taken place?

_____

_____

_____

8. How do we determine the amount of product formed in a chemical reaction or the amount of reactant required? How do we determine the limiting reagent? How do we determine percent yield?

_____

_____

_____

REACTIONS IN AQUEOUS SOLUTIONS

As you work through your AP Focus Study Guide, keep this chapter's Big Ideas in mind:

AP A LOOK AHEAD

BIG IDEA 1
The fundamental structure of the atom defines the ways by which atoms come together to form molecules.

BIG IDEA 2
The ways in which elements will interact with one another and with the external system can be predicted by their structures.

BIG IDEA 3
Elements will combine with one another in a predictable fashion that is described by the laws of physics.

BIG IDEA 6
Dissociation and recombination are opposing reactions, and the position that the reaction adopts is predicted by a discrete series of rules.

Section 4.1 General Properties of Aqueous Solutions, *pp. 119–121*

Essential Knowledge Covered:
2.A.1, 2.A.3, 2.D.1, 6.A.1

REVIEW IT

Using the word bank on the left, **fill in the blanks.**

Word Bank

aqueous
electrolyte
dissolve
nonconductive
nonelectrolyte
solute
solution
solvent
strong
water-based
weak

A homogeneous mixture of two or more substances is a _____; the substance that is present in a larger amount is the _____ and the substance present in a lesser amount is the _____. Any solution that has water as its solvent is referred to as _____, which literally means "_____." A substance that will form a solution with water is said to _____ in water. Any substance that will dissolve in water can be categorized into either an _____, which is a substance that, when dissolved in water, renders the resulting solution conductive to electricity, or a _____, which results in a _____ aqueous solution. Electrolytes can be _____, which means the entire sample dissociates, or _____, which means only a small portion of the sample is dissociated in solution.

USE IT

EK 2.D.1
Student Edition p. 120

Identify the strong electrolytes in the following group. What do all the strong electrolytes have in common?

CuCN	H_2SO_4	$TiCl_4$	CH_4
P_4S_{10}	$KClO_4$	H_2S	$AgCl2$
KOH	H_2O	$AgCl2$	$CsCO_3$

EK 2.A.1
Student Edition p. 119

Why are the vast majority of molecular compounds nonelectrolytes?

_____

_____

SUMMARIZE IT

EK 2.A.3, 6.A.1
*Student Edition
pp. 120–121*

Complete the following comparison charts:

Dissociates completely in water

An electrolyte
is

Does not dissociate completely in water

EK 2.A.3
*Student Edition
pp. 120–121*

A conductive
solution

Might contain a

Cannot contain only

Section 4.2   Precipitation Reactions, *pp. 121–126*

Essential Knowledge Covered:
3.A.1

USE IT

EK 3.A.1
Student Edition p. 122

Write the net ionic equation for the following reactions, using the solubility rules on page 122 of your textbook.

$Na_2S(aq) + ZnCl_2(aq) \rightarrow$

$(NH_4)_2CO_3(aq) + CaCl_2(aq) \rightarrow$

EK 3.A.1
Student Edition p. 122

Write the net ionic equation for the reaction between aqueous silver nitrate and aqueous sodium sulfate:

SUMMARIZE IT

EK 3.A.1

You have a mixture of gold(III) nitrate and silver(I) nitrate in an approximately 1:1 ratio. Knowing the prices of both gold and silver are at historical highs, you are very interested in separating the two metals. Suggest a chemical method by which this separation could be done, and write the full and net ionic equations for the reaction.

Section 4.3   Acid-Base Reactions, *pp. 126–132*

Essential Knowledge Covered:
1.E.1, 3.A.1, 3.B.2

EK 3.B.2
Student Edition p. 128

REVIEW IT

What effect does acid strength have on an acidic species' electrolyte strength? Why?

_____

_____

_____

EK 3.B.2
Student Edition p. 127

USE IT

Add the terms in the table below to the appropriate part of the Venn Diagram:

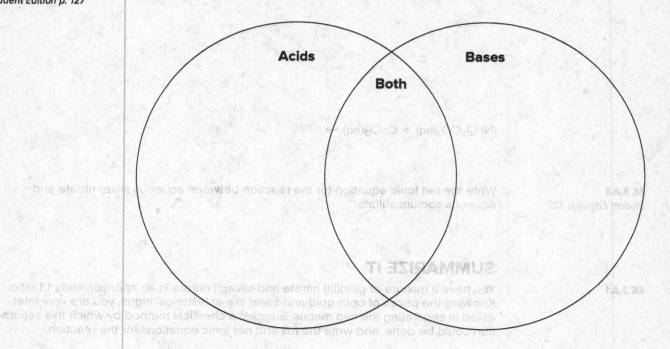

Taste sour	Taste bitter	Change litmus blue to red	Change litmus red to blue	React with metals to form H_2 gas
Have slippery feel	Electrolytic	React with carbonates to form CO_2 gas	Neutralized to form a salt and water	Proton acceptor
Proton donor	Forms H^+ ions	Forms HO^- ions	Defined by Arrhenius	Defined by Brønsted

EK 1.E.1, 3.A.1

SUMMARIZE IT

Write the molecular, ionic, and net ionic equations for each of the following acid-base reactions:

1. Potassium hydroxide + Hydrochloric acid

2. Hydrobromic acid + Magnesium hydroxide

3. Sulfuric acid + Sodium hydroxide

Essential Knowledge Covered:
1.E.1, 3.A.1, 3.B.1, 3.B.3

EK 3.B.3
Student Edition
pp. 132–134

Word Bank

Loss of electrons
gain of electrons
causes increase in
  oxidation state
causes decrease in
  oxidation state
caused by oxidizing
  agent
caused by reducing
  agent
half-reaction has
  electrons as products
half-reaction has
  electrons as reactants

REVIEW IT

Place the phrases from the word bank into the appropriate places on the cloud diagrams:

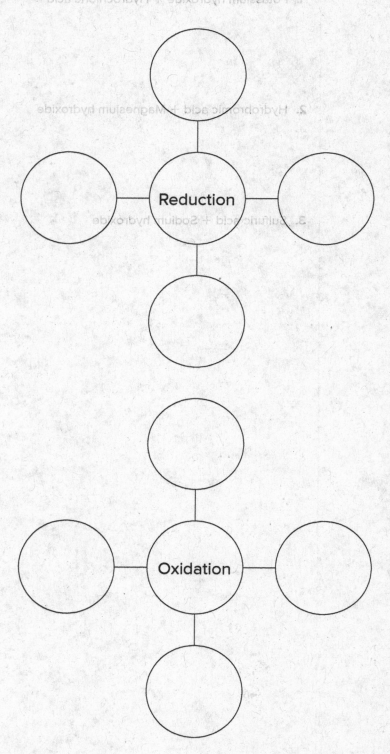

Section 4.4 Oxidation-Reduction Reactions (continued)

EK 3.B.3
Student Edition p. 136

USE IT

Using the rules on page 136 of your textbook, assign the oxidation number for each element in the following compounds or ions:

Cs_2O

_____

Al_2O_3

_____

$TiCl_4$

_____

EK 1.E.1, 3.A.1, 3.B.1, 3.B.3

SUMMARIZE IT

For the following reactions, identify whether or not it is a redox reaction. If so, (1) balance the reaction, (2) classify it according to type, (3) identify the oxidizing and reducing agents, and (4) write both half-reactions.

$Fe + O_2 \rightarrow Fe_2O_3$

$Cl_2 + NaBr \rightarrow NaCl + Br_2$

$SiCl_4 + NaF \rightarrow NaCl + SiF_4$

Section 4.5   Concentration of Solutions, *pp. 145–149*

Essential Knowledge Covered:
2.A.3

Vocabulary

Concentration of a
  solution
Molarity
Molar concentration
Dilution
Quantitative analysis

EK 2.A.3
Student Edition
pp. 145–146

EK 2.A.3
Student Edition
pp. 145–147

REVIEW IT

Using the word bank on the left, **fill in the blanks:**

The amount of solute present in a given amount of solvent is known as the

_____. This quantity is typically expressed in _____,

which is also known as _____ and is defined as the number of moles

of solute divided by the number of liters of solvent. Solutions of the same solute can be

prepared by _____, which is the addition of water to a solution of known

concentration to reduce that concentration by a known factor.

Determine the molarity of the following solutions:

0.6 moles KCl in 600 mL water.

117 g NaCl in 500 mL water.

USE IT

How would you prepare 30 liters of a stock solution that is 2.0 M in NaCl? How many
liters of 0.2 M NaCl solution could you prepare via dilution of this stock solution?

EK 2.A.3
Student Edition
pp. 145–147

SUMMARIZE IT

You need to make a very dilute solution for a particular biochemical assay. To do this, your supervisor gives you the following set of directions:

1. Dissolve 0.88 g glucose in 500 mL water.
2. Remove 2.0 mL of this solution and dilute it to 1000 mL.
3. Remove 10 mL of this solution and dilute it to 250 mL.

What is the final molarity of the glucose solution? Why can't a solution of this molarity be prepared directly?

Section 4.6   Gravimetric Analysis, *pp. 149–151*

Essential Knowledge Covered:
1.A.2, 1.E.2, 3.A.1, 3.A.2

EK 1.A.2, 1.E.2
Student Edition p. 149

REVIEW IT

What is gravimetric analysis, and why is it considered a quantitative analytical technique?

_____

_____

_____

Section 4.6 Gravimetric Analysis (continued)

SUMMARIZE IT

EK 1.E.2

Could gravimetric analysis be used to determine the % Cl from the following reaction? Why or why not?

$NaCl(aq) + KF(aq) \rightarrow NaF(aq) + KCl(aq)$

_____

_____

EK 3.A.1, 3.A.2

A sample of 0.6760 g of an aqueous Barium-containing compound is reacted with excess Na_2SO_4. If the mass of the $BaSO_4$ formed is 0.4105 g, then what was the percent by mass of Ba in the original sample.

Section 4.7 Acid-Base Titrations, *pp. 151–154*

Essential Knowledge Covered:
3.A.2, 3.B.2

REVIEW IT

EK 3.A.2
Student Edition
pp. 151–153

How many grams of KHP are needed to neutralize 18.64 mL of a 0.1004 M NaOH solution?

USE IT

EK 3.A.2, 3.B.2
Student Edition
pp. 151–153

How does the pH at the equivalence point of a titration relate to the pH at which point the indicator changes color?

_____

_____

Section 4.7   Acid-Base Titrations (continued)

Based on your answer to the above question, select the best indicator from the table (below) for a reaction whose equivalence point is expected to be between pH 4.5 and 5.0:

Substance	Colour change	Colour change pH range
Thymol blue	Red – yellow	1.2 – 2.8
Methyl orange	Red – orange	3.1 – 4.4
Methyl red	Red – yellow	4.4 – 6.2
Lithmus	Red – blue	5.0 – 8.0
Bromthymol blue	Yellow – blue	6.0 – 7.6
Phenol phthalein	Colourless – red	8.2 – 9.8
Thymol phthalein	Colourless – blue	9.3 – 10.5

SUMMARIZE IT

EK 3.A.2, 3.B.2

What is the concentration of an HCl solution if 10.00 mL are added to 60.2 mL of a 0.427 M KOH solution, at which point the indicator changes color?

Section 4.8   Redox Titrations, *pp. 155–157*

Essential Knowledge Covered:
1.A.2, 1.E.2, 3.A.2, 3.B.3

REVIEW IT

EK 3.A.2, 3.B.3
Student Edition
pp. 155–156

How does a Redox titration compare to a normal (acid-base) titration? How is it different?

SUMMARIZE IT

EK 1.A.2, 1.E.2, 3.A.2

Redox titrations commonly use internal indicators such as $KMnO_4$ or $Na_2Cr_2O_7$, since these compounds change color based on their oxidation states, but can use external indicators (such as phenolphthalein) as in acid-base titrations. Based on the reaction type, place a check mark in the box corresponding to the best choice of indicator.

Half Rxn	$MnO_4^- \rightleftharpoons Mn^{2+}$	phenolphthalein
$NH_3 \rightarrow NH_4OH$		
$OH^- \rightarrow H2O$		
$H^+ \rightarrow H2O$		
$Fe^0 \rightarrow Fe_2O_3$		
$NaBr \rightarrow Br_2$		

AP Reviewing the Essential Questions

These questions were posed in the *Chemistry* chapter opener (page 118). Answer them using the knowledge you've gained from this chapter.

1. Why is water an effective solvent for ionic compounds?

_____

_____

_____

2. What is a reversible reaction?

_____

_____

_____

3. What is an acid, and what is a base? How can we determine an acid or base's concentration?

_____

_____

_____

4. How do we represent an acid-base reaction? How do we represent an oxidation-reduction reaction?

_____

_____

GASES

As you work through your AP Focus Review Guide, keep this chapter's Big Ideas in mind:

AP A LOOK AHEAD

BIG IDEA 1 Gas behavior changes in response to changes in volume, temperature, pressure, and concentration.

BIG IDEA 2 Forces between particles drive a gas's response to changes in temperature, volume, and other factors.

BIG IDEA 3 Gas behavior during a chemical reaction can be understood through balanced stoichiometric equations.

BIG IDEA 5 Gas behavior in response to temperature can be understood through the Kinetic Molecular Theory.

Section 5.1   Substances That Exist as Gases, *pp. 173–174*

Essential Knowledge Covered:
2.B.3

REVIEW IT

Using the word bank on the left, **fill in the blanks.**

Word Bank

allotrope
diatomic
molecular
monatomic
nitrogen
oxygen

EK 2.B.3

Our atmosphere is made up of roughly 78% _____, 21% _____ and a mixture of many other gases make up the remaining 1%. Gases such as nitrogen, hydrogen, fluorine and chlorine all exist as _____ gases while gases called the noble gases exist as _____ gases. There is a(n) _____ of oxygen called ozone that also exists as a gas at room temperature. Many other gases exist, but these are all in _____ form.

SUMMARIZE IT

What is the difference between the term gas and the term vapor?

_____

_____

Section 5.2   Pressure of a Gas, *pp. 174–178*

Essential Knowledge Covered:
2.A.2

REVIEW IT

Using the word bank on the left, **fill in the blanks.** Words can be used more than once.

Word Bank

atmospheric pressure
barometer
manometer
newton
pressure
pascal
standard atmospheric
   pressure

When force is measured in _____ and area in m^2, then pressure is defined as the force per m^2, or N/m^2. This unit is also called the _____. The gases in our atmosphere exert downward force on the Earth called _____. This pressure is measured using a _____. The simplest of these involves a column of mercury contained in a closed tube inverted in a reservoir of mercury. When the column of mercury measures 76 cm at sea level and at 0°C, the pressure is said to be at _____.

Gases other than the atmosphere have their pressure measured using a _____. Two types of these exist. A closed tube setup is usually used to measure pressures below _____ and an open tube design is used to measure higher pressures.

USE IT

EK 2.A.2
Student Edition pp. 176–177

Fill in the operation that must be carried out in the conversions indicated in the following flow charts.

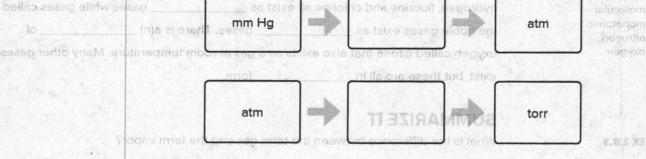

EK 2.A.2
Student Edition p. 177

Use the flow charts from above to put the pressures given here in order from lowest to highest by converting all measures to the same unit.

900 mm Hg, 1.066 atm, 770 mm Hg, 96.66 kPa, 1.138 atm

Section 5.2   Pressure of a Gas (continued)

Section 5.3   The Gas Laws (continued)

SUMMARIZE IT

EK 2.A.2

In the barometer on page 176 of your textbook, the column height is given as 76 cm of mercury. To what conditions could this correspond?

_____

_____

Explain what would happen to the column height if the atmospheric pressure increases.

_____

_____

_____

Section 5.3   The Gas Laws, *pp. 178–184*

Essential Knowledge Covered:
1.E.1, 2.A.2

USE IT

EK 2.A.2
Student Edition pp. 182–183

Match the law to its corresponding proportionality and to the variables that must remain constant to the right. Include the relationship of the law in the last column as either direct or inverse.

Proportionality

$V \propto \dfrac{1}{P}$

$V \propto T$

$V \propto n$

Constants

pressure
temperature
volume
amount of gas

EK 1.E.1, 2.A.2
Student Edition pp. 182–183

Proportionality	Law	Constant	Relationship
	Avogadro's law		
	Boyle's law		
	Charles's law		

Explain the effect on the volume for each of the following changes:

At a constant temperature and pressure, the amount of gas in a sample triples

_____

_____

At a constant amount of gas and pressure, the Kelvin temperature of a gas is decreased by a factor of 0.5

_____

_____

_____

At a constant amount of gas and temperature, the pressure on the gas sample doubles.

_____

_____

_____

Section 5.3   The Gas Laws (continued)

SUMMARIZE IT

EK 1.E.1, 2.A.2

Complete the following graphs for the indicated laws.

Boyle's law

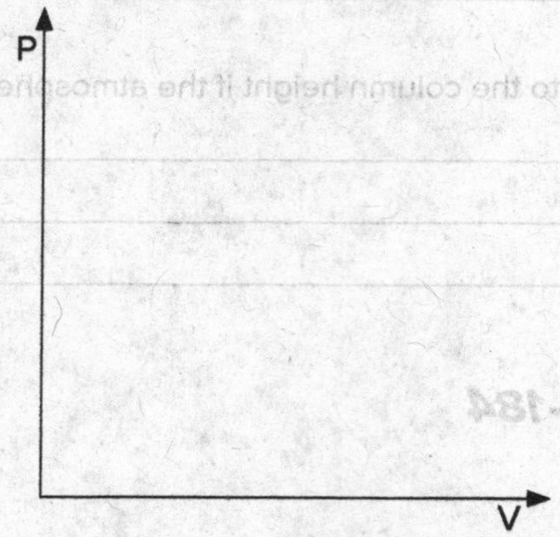

P

V

Charles's law

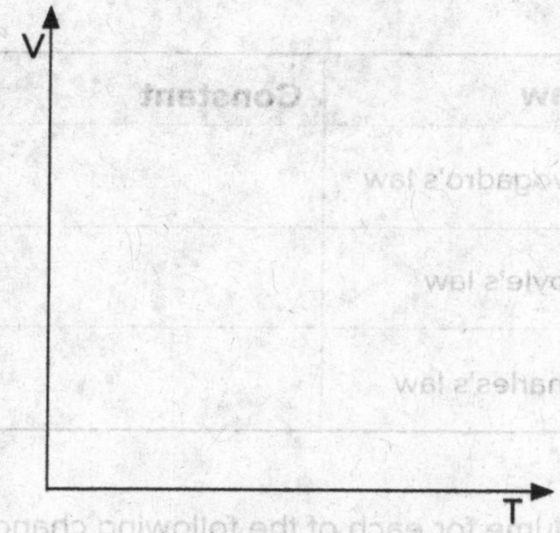

V

T

EK 1.E.1, 2.A.2

Does a gas undergo the same change in volume between 300K to 600K as it does when going from 300°C to 600°C? Why or why not?

Section 5.4   The Ideal Gas Equation, *pp. 184–193*

Essential Knowledge Covered:
2.A.2

REVIEW IT

EK 2.A.2
Student Edition pp. 184–185

What are the two assumptions made for a gas to be an ideal gas?

_____

_____

Explain how this is supported or not supported by Coulomb's law.

_____

_____

_____

Why are these assumptions valid for most gases under 'normal" conditions?

_____

_____

_____

USE IT

EK 2.A.2
Student Edition pp. 185–186

These two triangles can be used to take the Ideal gas equation and rearrange it for any of the variables in the equation.

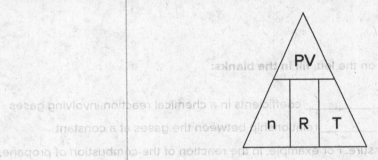

The triangles work by covering up the quantity that you are solving for, and the other terms then show the formula. Anything that is above the horizontal line is in a numerator position and anything below the horizontal line is in a denominator position. An example to solve for *n* is given below:

With the *n* covered up, the formula becomes $n = \dfrac{PV}{RT}$

Use these triangles to write the formulae for the other variables.

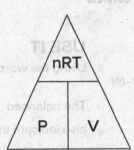

Section 5.4   The Ideal Gas Equation (continued)

EK 2.A.2
Student Edition p. 186

A sample of gas at 75.0°C is held in a 2.15 L container at 11.75 atm. How many moles of gas are in the container under these conditions?

_____

_____

_____

SUMMARIZE IT

EK 2.A.2

A sample of water vapor forms when 8.50 g of water is heated to 435°C in a 3.75 L container. Find the pressure of the water vapor (in atm) from this given information.

_____

_____

_____

_____

Section 5.5   Gas Stoichiometry, *pp. 193–195*

Essential Knowledge Covered:
2.A.1, 2.A.2, 3.A.2

USE IT

EK 2.A.2, 3.A.2
Student Edition pp. 193–194

Using the word bank on the left, **fill in the blanks:**

Word Bank

five
four
stoichiometric
three
volume

The balanced _____ coefficients in a chemical reaction involving gases also indicate the _____ relationship between the gases at a constant temperature and pressure. For example, in the reaction of the combustion of propane, $C_3H_8(g) + 5 O_2(g) \rightarrow 3 CO_2(g) + 4 H_2O(g)$, for every liter of propane that reacts, there will be _____ liters of carbon dioxide gas and _____ liters of water vapor released. The reaction also tells us that _____ liters of oxygen gas are needed in the combustion process. This all assumes that the temperature and pressure do not change during the reaction.

SUMMARIZE IT

EK 2.A.1, 2.A.2, 3.A.2

The reaction for the decomposition of hydrogen peroxide is

$H_2O_2(l) \rightarrow H_2O(l) + O_2(g)$

Balance the chemical reaction.

Section 5.5   Gas Stoichiometry (continued)

EK 2.A.1, 2.A.2

If 77.20 g of hydrogen peroxide decomposes, how many grams of water will be formed?

At 28.0°C and 854 mm Hg, what volume of oxygen gas will form from the reaction above?

Section 5.6   Dalton's Law of Partial Pressures, *pp. 195–202*

Essential Knowledge Covered:
2.A.2

USE IT

EK 2.A.2
Student Edition p. 197

Below is a diagram of a combination of gases. The circles represents Gas A, the squares represents Gas B and the triangles represents Gas C. The total pressure of the system is 320 mm Hg. If the number of each shape in the diagram represents the relative abundance of each gas in the mixture, determine the partial pressure of each gas in the mixture **without using your calculator.**

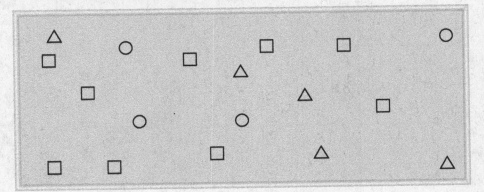

Section 5.6   Dalton's Law of Partial Pressures (continued)

EK 2.A.2
Student Edition p. 198

In a mixture of gases, there are 3.85 moles of carbon dioxide gas, 1.75 moles of oxygen gas and 1.55 moles of carbon monoxide gas. If the total pressure is 8.25 atm, what would be the partial pressure of each gas? Before you start your calculations, write out the steps/conversions necessary to find the final answer.

SUMMARIZE IT

EK 2.A.2

A gas is collected by the downward displacement of water at 30°C. Once collected, the pressure of the container holding the gas was 875.30 mm Hg. Using the table below, determine the pressure of the collected gas in the container.

Water Vapor Pressures	
Temp (°C)	Pressure (mm Hg)
10	9.21
20	17.54
30	31.82
40	55.32
50	92.51
60	149.38

Section 5.7   The Kinetic Molecular Theory of Gases, *pp. 202–210*

Essential Knowledge Covered:
2.A.2, 5.A.1

EK 2.A.2, 5.A.1
Student Edition p. 203

REVIEW IT

What are the four assumptions of the kinetic theory of gases?

_____

_____

_____

_____

_____

_____

_____

_____

USE IT

EK 2.A.2, 5.A.1
Student Edition pp. 203–204

Illustrate how the kinetic theory of gases can explain why gases can be compressed. Include an explanation after your diagrams.

EK 2.A.2, 5.A.1
Student Edition pp. 203–204

Use the kinetic theory of gases to explain each observation below:

As the temperature of a balloon filled with helium decreases, the volume of the balloon decreases.

_____

_____

_____

A basketball goes flat as air escapes through a small hole in the ball.

_____

_____

_____

On a cold day, a truck driver puts air into a low tire, but after many hours of travel, the tire pressure gets so high that the tire ruptures.

_____

_____

_____

SUMMARIZE IT

EK 2.A.2, 5.A.1

Since the average kinetic energies of two samples of gases are equal at the same temperature, what can be concluded about the two gases at the same temperatures shown on this graph?

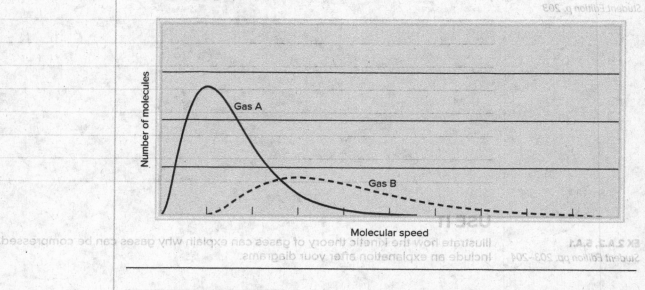

_____

_____

_____

Section 5.8   Deviation from Ideal Behavior, *pp. 210–213*

Essential Knowledge Covered:
2.A.2

REVIEW IT

EK 2.A.2
Student Edition pp. 210–211

Using the word bank on the left, **fill in the blanks**

Word Bank

attraction
attractive forces
condense
decrease
ideal behavior
non-ideal behavior
pressure
temperature
van der Waals
volume

When real gases experience changes that cause particles to get close together,

_____ can be experienced. This is the case whenever a gas does not

follow $PV = nRT$, which would be _____. Effects of this include gas particles

experiencing _____ as they are forced closer and closer together by

either an increase in _____ or a decrease in _____. As the _____

is lowered, the gas particles will _____ into a liquid due to the _____

between the particles. _____ equation attempts to correct ideal behavior to

non-ideal behavior by introducing a correction factor. Other corrections can be intro-

duced, such as a correction for the volume that the gas particles occupy. This would

_____ the actual volume of empty space that the gas is free to move through.

USE IT

EK 2.A.2
Student Edition pp. 212–213

The Van der Walls equation is $\left(P + \dfrac{an^2}{V^2}\right)(V - nb) = nRT$

The table to the left gives vales for a and b for various gases.

Gas	a	b
Neon	0.211	0.0171
Nitrogen	1.39	0.0391
Ammonia	4.17	0.0371

At 53°C, a sample of $N_2(g)$ containing 5.21 moles of gas is held in a 2.85 L container.

Find the ideal gas pressure and the non-ideal pressure of the system, in atm.

Why is the non-ideal pressure lower than the ideal pressure?

_____

_____

_____

AP Reviewing the Essential Questions

These questions were posed in the *Chemistry* chapter opener (page 172). Answer them using the knowledge you've gained from this chapter.

1. What are the characteristics of a gas?
_____
_____

2. What are intermolecular forces?
_____
_____
_____
_____

3. How are pressure, temperature, volume, and moles of gas interrelated?
_____
_____
_____
_____

4. Is the mole-volume relationship conserved in a chemical reaction?
_____
_____
_____
_____

5. How do we calculate volumes of gas produced or required in a chemical reaction?
_____
_____
_____
_____

6. How do we determine the molecular mass of an unknown gas?
_____
_____
_____

7. How is the kinetic energy of molecules related to temperature?
_____
_____
_____

Chapter 6

THERMOCHEMISTRY

As you work through your AP Focus Review Guide, keep this chapter's Big Ideas in mind:

AP — A LOOK AHEAD

BIG IDEA 1 — Adjacent atoms share electrons in the form of a bond which requires energy to form and releases energy when broken.

BIG IDEA 3 — Chemical reactions include heat energy as either a product or reactant; the entire system and its surroundings must be evaluated to determine which.

BIG IDEA 5 — The energy stored in the chemical bond is often released as heat when the bond is broken.

Section 6.1   The Nature of Energy and Types of Energy, *p. 231*

Essential Knowledge Covered:
5.B.1, 5.B.2, 5.C.1, 5.D.1

Vocabulary

energy
radiant energy
thermal energy
chemical energy
potential energy

REVIEW IT

Using the vocabulary on the left, **fill in the circles** relating to the hierarchy of energy types, then use the completed diagram to fill in the blanks in the next sentence.

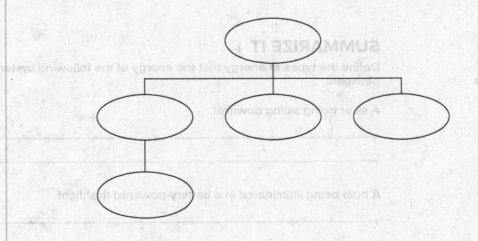

EK 5.C.1, 5.D.1
Student Edition p. 231

From this diagram, _____ is considered a subset of _____.
Explain this categorization.

_____

_____

_____

USE IT

EK 5.B.2
Student Edition p. 231

The theoretical pendulum shown below swings through frictionless air. **Label** the diagram to indicate the types of energy that are being interconverted when the pendulum swings from A to B to C.

A   B   C

SUMMARIZE IT

EK 5.B.2

Define the types of energy that the energy of the following systems is being converted between:

A skier going skiing downhill.

_____

_____

A bulb being illuminated in a battery-powered flashlight.

_____

_____

Driftwood burning in a bonfire.

_____

_____

EK 5.B.1

Summarize the Law of Conservation of Energy.

_____

_____

_____

Section 6.2 Energy Changes in Chemical Reactions, *pp. 232–233*

Essential Knowledge Covered:
3.C.2, 5.B.1, 5.B.2, 5.B.3,
5.C.2

REVIEW IT

EK 5.B.1, 5.B.2
Student Edition p. 232

Place an 'X' in the boxes above the transitions that are not permitted by each type of system, then give an example of that type of system:

☐ ☐   ☐ ☐   ☐ ☐
Mass Energy   **Mass Energy**   **Mass Energy**
↕   ↕   ↕   ↕   ↕   ↕
Isolated System   Closed System   Open System

Isolated System: _____

Closed System: _____

Open System: _____

SUMMARIZE IT

EK 3.C.2

Is it possible to assess the exo- or endothermicity of an isolated system? Why or why not?

_____

_____

EK 5.B.3, 5.C.2

The dissolution of potassium carbonate in water is an endothermic process. Would this dissolution occur more readily with the surroundings at a low temperature or at a high temperature? Why?

_____

_____

_____

_____

EK 5.C.2

Why are decomposition reactions usually exothermic, whereas combination reactions are usually endothermic?

_____

_____

_____

_____

Section 6.3 Introduction to Thermodynamics, *pp. 234–239*

Essential Knowledge Covered:
3.C.2, 5.B.1, 5.B.2, 5.B.3

USE IT

EK 5.B.1, 5.B.2
Student Edition p. 236

Complete the table of sign conventions for work and heat, as they relate to the system.

Process	Sign
work done by the system on the surroundings	
work done on the system by the surroundings	
heat absorbed by the system from the surroundings	
heat released by the system to the surroundings	

EK 3.C.2, 5.B.1
Student Edition p. 236

Justify your choice of sign for 'heat release by the system to the surroundings' in the table above:

_____

_____

_____

EK 5.B.1, 5.B.2
Student Edition p. 238

In a piston, the work done when the cylinder is compressed from 4 L to 2 L is 373 J. During this compression, there are 97 J of heat energy transferred to the surroundings. What is the change in internal energy of the system?

SUMMARIZE IT

EK 5.B.1, 5.B.2, 5.B.3

Consider the following chemical reactions occurring at constant pressure within the chamber to the left. Determine if work is done by the system, on the system, or if no work is done.

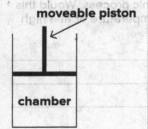

$2O_3 (g) \rightarrow 3O_2 (g)$

_____

_____

_____

$H_2 (g) + Cl_2 (g) \rightarrow 2 HCl (g)$

_____

_____

$Br_2 (g) \rightarrow Br_2 (l)$

_____

_____

_____

Section 6.4 Enthalpy of Chemical Reactions, *pp. 240–246*

Essential Knowledge Covered:
1.E.2, 3.C.2, 5.B.2,

Word and Equation Bank:

enthalpy
enthalpy of reaction
thermochemical
    equation
$H = U + PV$
$\Delta H = H_{prod} - H_{react}$
$\Delta H > 0$
$\Delta H < 0$
$\Delta U = \Delta H - RT\Delta n$

EK 1.E.2, 3.C.2, 5.B.2
Student Edition
pp. 245–246

REVIEW IT

Using the vocabulary and equations on the left, **fill in the blanks.**

The thermodynamic state of a system is described by the state function called

_____, which is abbreviated H and defined by the equation _____.

Since most reactions occur at constant pressure, the change in systemic enthalpy is

calculated by the difference between the initial and final states, or symbolically as

_____. This change in enthalpy for a reaction is referred to as the

_____ . The value of ΔH is _____ for an exothermic reac-

tion and _____ for an endothermic reaction. When the value for ΔH is

included with the mass relationships in an equation, the resultant equation is termed a

_____, because it contains both thermodynamic and chemical

information. ΔH is a component of the overall energetics of a system, and is used to

calculate ΔU via the equation _____.

USE IT

Consider the equation:

$P_4(s) + 5O_2(g) \rightarrow P_4O_{10}(s) \quad \Delta H = -3013 \text{ kJ/mol}$

Is this reaction exothermic or endothermic? How do you know?

_____

_____

Calculate the heat produced when 300.0 g P_4 is burned in air.

Calculate ΔU for the system if 1.0 mol P_4 is burned in air at 25 °C.

Why does ΔH differ from ΔU in the burning of P_4?

_____

SUMMARIZE IT

EK 3.C.2, 5.B.2   Consider the diagram below:

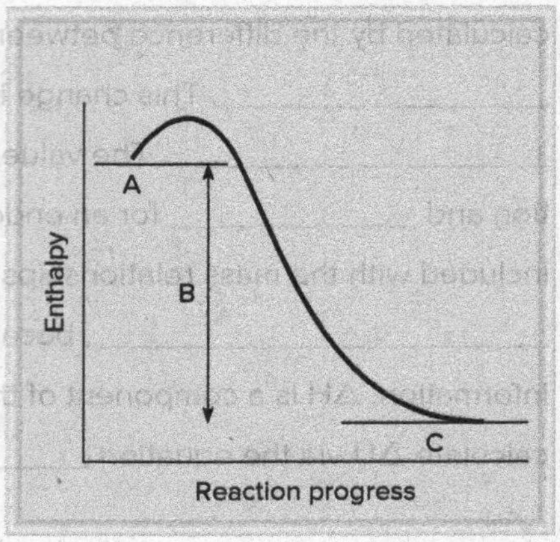

What label should be at position A?

_____

What label should be at position C?

_____

What does distance B represent?

_____

What type of reaction does this represent, from a thermodynamic perspective?

_____

Which of the following reactions could correspond to this energy diagram:

    a. CH_4 (g) + $2O_2$ (g) → CO_2 (g) + $2H_2O$ (g)
    b. H_2O (s) → H_2O (l)
    c. H_2O (l) → H_2O (g)
    d. CO_2 (g) + $2H_2O$ (g) → CH_4 (g) + $2O_2$ (g)

How does ΔH compare with ΔU for this reaction?

_____

EK 5.B.1, 5.B.2, 5.B.4
Student Edition
pp. 251–253

USE IT

If a constant pressure calorimeter contained 100.00 mL water at 23.000 °C before a 30.000 g aluminum ingot at 100.000 °C was placed inside the device, what temperature would the water reach upon complete equilibration? The specific heat of aluminum is 0.900 J/g °C.

EK 5.B.1, 5.B.2, 5.B.3, 5.B.4

SUMMARIZE IT

A 12.18 g sample of a metal is heated to 65.00 °C and then added to a constant pressure calorimeter that contains 25.00 g water at 25.55 °C. The temperature increases to 27.25 °C. Based on the table below, what is the metal?

Metal	Specific Heat (J/g °C)
Aluminum	0.900
Iron	0.450
Copper	0.387
Lead	0.158
Mercury	0.139
Gold	0.129

Word Bank

standard enthalpy of formation

standard enthalpy of reaction

standard state

Hess's Law

$\Delta H°_{rxn} = \Sigma n \Delta H°_f(\text{products}) - \Sigma m \Delta H°_f(\text{reactants})$

EK 5.C.2
Student Edition
pp. 256–259

REVIEW IT

In order to provide a uniform starting point for enthalpy calculations, the _____ _____ for any element in its most stable form is zero. To be in the _____, an element must be at 1 atm of pressure. Due to the fact that a standardized point of reference has been established, the _____ _____ can be calculated for a reaction wherein a compound is synthesized from its component elements at 1 atm using the equation _____ _____. If a compound cannot be synthesized from its component elements, of which there are many examples, then their $\Delta H°_f$ can be determined using _____, which states that the enthalpy change of an overall process is the sum of the enthalpy changes of its individual steps.

SUMMARIZE IT

Calculate ΔH for the following reaction:

$CO (g) + NO (g) \rightarrow CO_2 (g) + \frac{1}{2} N_2 (g)$ $\Delta H = ?$

Given the following thermochemical equations:

a. $CO (g) + \frac{1}{2} O_2 \rightarrow CO_2 (g)$ $\Delta H = -283.0 \text{ kJ}$
b. $N_2 (g) + O_2 (g) \rightarrow 2NO (g)$ $\Delta H = 180.6 \text{ kJ}$

Essential Knowledge Covered:
5.B.3, 5.C.2

Vocabulary

heat of solution
heat of hydration
heat of dilution
lattice energy

REVIEW IT

In the diagram below, identify the state function that defines the energy change at A-D, using the vocabulary on the left, then give a definition of each state function:

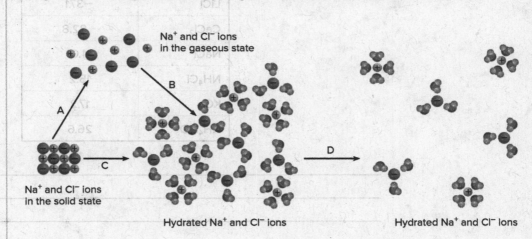

Na$^+$ and Cl$^-$ ions in the gaseous state

Na$^+$ and Cl$^-$ ions in the solid state

Hydrated Na$^+$ and Cl$^-$ ions

Hydrated Na$^+$ and Cl$^-$ ions

SUMMARIZE IT

EK 5.B.3, 5.C.2
Student Edition p. 261

You are designing a set of hold and cold packs to be used for first aid kits. The packs consist of an inner vial that contains a compound suspended in an outer packet that is full of water, as shown below. When the inner vial is broken, the compound becomes rapidly solvated.

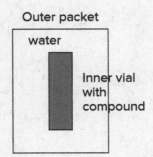

Outer packet

water

Inner vial
with
compound

Section 6.7 Heat of Solution and Dilution (continued)

Select the best compound for both the hot pack and the cold pack from the table below, and explain your choices:

Compound	ΔH_{soln} (kJ/mol)
LiCl	−37.1
CaCl$_2$	−82.8
NaCl	4.0
NH$_4$Cl	15.2
KCl	17.2
NH$_4$NO$_3$	26.6

_____

_____

_____

_____

These questions were posed in the *Chemistry* chapter opener (page 230). Answer them using the knowledge you've gained from this chapter.

1. What is chemical potential energy of atoms?

_____

_____

2. What is the chemical potential energy of molecules?

_____

_____

3. What is heat?

_____

_____

4. What is the difference between an exothermic and an endothermic reaction?

_____

_____

5. Does adding heat to a system change its energy content? What is the direction of energy flow in a chemical reaction?

_____

_____

6. What is the relationship between energy, work, and heat?

_____

_____

7. How can a balanced chemical equation be used to determine heats of reaction?

_____

_____

Chapter 7

QUANTUM THEORY AND THE ELECTRONIC STRUCTURE OF ATOMS

As you work through your AP Focus Review Guide, keep this chapter's Big Ideas in mind:

AP A LOOK AHEAD

BIG IDEA 1   Quantum theory provides the best current model for understanding the strucuture of the atom

Section 7.1   From Classical Physics to Quantum Theory, *pp. 275–279*

Essential Knowledge Covered:
1.C.2

REVIEW IT

EK 1.C.2
Student Edition pp. 275–277

Word Bank

amplitude
continuous
electromagnetic
frequency
light
quanta
quantum theory
quantum
radiation
wave
wavelength

Using the word bank on the left, **fill in the blanks.** Words may be used more than once.

Classical physics failed to properly explain how atoms could be stable. A new physics would be needed to study sub-atomic particles. Max Planck discovered that energy was emitted by atoms and molecules in discrete, not _____ quantities. These packets of energy were referred to as _____. He suggested that each packet of energy was released in a _____ process. Planck's _____ lead to the development of many new branches of physics.

To understand this new theory, it is necessary to understand the properties of a _____. These have measureable characteristics. The _____ is the distance between identical points on successive waves. The _____ is the number of waves that pass through a stationary point every second and the _____ is the vertical distance from the midline of a wave to a peak or trough. The speed of a wave is found by the product of the _____ and the _____.

Later, James Maxwell proposed that a(n) _____ wave, carrying both an electric field component and a magnetic field component. They travel at the same speeds, but are travelling in perpendicular planes to each other. The speed at which these waves travel is the speed of _____. Planck gave the name _____ to the smallest quantity of energy that can be emitted or absorbed in the form of _____ radiation.

USE IT

EK 1.C.2
Student Edition p. 277

What is the frequency of light that travels at a wavelength of 497 nm?

What is the wavelength of an electromagnetic wave that travels at a frequency of 6.25×10^{14} Hz?

SUMMARIZE IT

EK 1.C.2

For the wave given here, label the wavelength and amplitude of the wave.

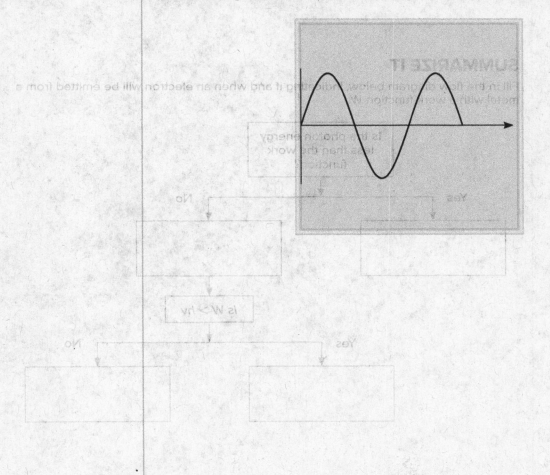

Section 7.2   The Photoelectric Effect, *pp. 279–281*

Essential Knowledge Covered:
1.B.1, 1.D.3

USE IT

EK 1.B.1, 1.D.3
Student Edition p. 280

A photon has a wavelength of 7.35×10^4 nm. Use the equation $E = h\dfrac{c}{\lambda}$ to find the energy of this photon.

EK 1.B.1, 1.D.3
Student Edition p. 281

A metal has a work function of 4.15×10^{-19} J. What is the minimum frequency of light required to release electrons from the surface of the metal?

SUMMARIZE IT

EK 1.B.1, 1.D.3

Fill in the flow diagram below, indicating if and when an electron will be emitted from a metal with a work function *W*.

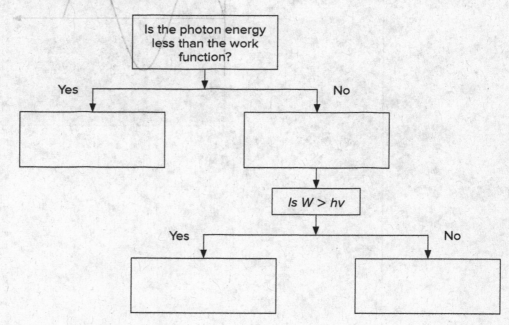

Essential Knowledge Covered:
1.C.2, 1.D.3

REVIEW IT

Using the word bank on the left, **fill in the blanks.**

Word Bank

continuous
emission spectra
excited state
ground state
line
photon
quantized
wavelengths

For centuries, scientists had known that light from the sun was composed of all colors

of the spectrum. However, when studying the light emitted by atoms, the spectra

emitted was not _____, in fact it was made of only specific colors termed the

_____ for the element. This is also called a _____ spectra, based on

the fact that only light of specific _____ can be seen in the spectra from

these elements.

When energy of a large enough quantity is added to the atoms of an element, the

electrons move from the _____ to a(n) _____. This energy is then

released as a _____ of that particular energy, corresponding to a spectral line for

the element. An electron can only move between energy levels if energy of that

specific or _____ amount is available to promote the electron.

USE IT

EK 1.C.2
Student Edition p. 284

According to the laws of classical physics, what should the electron of a hydrogen
atom experience when no energy is added to the system?

_____

_____

EK 1.C.2, 1.D.3
Student Edition p. 285

When is the energy associated with $\Delta E = R_H \left(\dfrac{1}{n_i^2} - \dfrac{1}{n_f^2} \right)$ positive and when is it negative?

_____

_____

_____

Section 7.3   Bohr's Theory of the Hydrogen Atom (continued)

SUMMARIZE IT

EK 1.C.2

Analogies often help organize our ideas of scientific theories and laws. Develop an analogy for the concept of the electron in a hydrogen atom, first classically, and then using the concept of the quantized atom.

_____

_____

_____

_____

_____

Section 7.4   The Dual Nature of the Electron, *pp. 287–291*

Essential Knowledge Covered:
1.C.2, 1.D.3

REVIEW IT

EK 1.C.2
Student Edition p. 290

Find Figure 7.13b on page 290 of your textbook. Explain why the diagram does not fit with the concept associated with de Broglie.

_____

_____

How would you modify Figure 7.13b to match the concept presented by de Broglie? Include a sketch of this modification.

USE IT

EK 1.C.2, 1.D.3
Student Edition p. 290

De Broglie deduced that the particle and wave properties are related by the expression $\lambda = \dfrac{h}{mu}$, where λ, m, and u represent wavelength, mass and velocity respectively. If a photon has a wavelength of 4.52×10^{-9} m, what would be the mass of the photon?

Section 7.4   The Dual Nature of the Electron (continued)

EK 1.D.3

$$\lambda = \frac{h}{mu}$$

where
λ is the wavelength,
m is the mass of the particle
and u is its velocity.

SUMMARIZE IT

Calculate the wavelength of an electron moving at 4.25×10^7 m/s, using the equation $\lambda = \dfrac{h}{mu}$ and comment on the magnitude of the answer.

Section 7.5   Quantum Mechanics, *pp. 291–295*

Essential Knowledge Covered:
1.C.2, 1.D.1

REVIEW IT

Using the word bank on the left **fill in the blanks.** Words may be used more than once.

Word Bank

atomic orbital
electron density
emission spectra
Heisenberg
momentum
probability
position
quantum mechanics
Schrodinger wave
   equation
wave function

The _____ uncertainty principle states that it is impossible to simultaneously

know the _____ and _____ of a particle with certainty. As a result, the

quantized electron concept suggested by Bohr to explain the _____

coupled with the _____ uncertainty principle lead to a new physics.

The _____ incorporates both particle behavior in terms of

mass and wave behavior in terms of a _____ ψ which depends on the

location in space of the electron. The square of this function relates to the

_____ of finding an electron in a region of space. This equation started a new

era in physics in the field called _____. This field tells us that we

cannot pinpoint the _____ of an electron in an atom, but it gives us the

_____ that an electron can be found in a region of space known as the

_____. In quantum mechanics, we discuss a(n) _____, rather

than the orbit of an electron. It can be thought of as the _____ of an

electron in an atom.

USE IT

EK 1.C.2, 1.D.1
Student Edition p. 292

What is the equation associated with the Heisenberg uncertainty principle? Explain the meaning of the terms in the expression that are not constants.

_____

_____

_____

How does this principle affect our thought of how electrons move as they orbit a nucleus?

_____

_____

SUMMARIZE IT

EK 1.C.2, 1.D.1

Fill in the timeline chart with the theories as the electron in an atom was improved.

EK 1.C.2, 1.D.1

$$\Delta x \Delta p \geq \frac{h}{4\pi}$$

The estimated uncertainty in measuring the speed of an electron is 1.7%. If an electron is moving at 3.5×10^6 m/s, what is the uncertainty in measuring the position of the electron?

Section 7.6   Quantum Numbers, *pp. 295–297*

Essential Knowledge Covered:
1.B.2, 1.C.2

EK 1.B.2, 1.C.2
Student Edition
pp. 295–296

USE IT

For the given values of *l*, give the name of the orbital in the following chart:

l	Name of Orbital
0	
3	
1	
2	

What are the origins of the letters s, p and d?

_____

_____

SUMMARIZE IT

EK 1.B.2, 1.C.2

While there are four quantum numbers, only three are required to describe the electron in hydrogen. Why is this?

_____

_____

_____

Essential Knowledge Covered:
1.B.2

REVIEW IT

Using the word bank on the left, **fill in the blanks.** Words may be used more than once.

Word Bank

boundary surface
   diagram
d-orbital
electron density
energy
orbital(s)
orientation
p-orbital
s-orbital
shape

The number of values of magnetic quantum number determines the number of

_____. For example, if $l = 0$, then $(2l + 1) = 1$, so there is only one _____,

which corresponds to the _____. When $l = 1$, then $(2l + 1)$ gives the number of

_____, so $2(1) + 1 = 3$. This is associated with the three _____ that exist

for multi-electron atoms. When $l = 2$, then $(2l + 1) = 5$, which corresponds to the five

_____ that exist.

The _____ is a representative drawing of the shape

corresponding to the _____ for a particular orbital. For an s-orbital, this

diagram is simply a sphere. The only difference between 1s and 2s is the size of the

sphere that would be drawn. The three diagrams for p orbitals are identical in

_____ and _____, only their _____ differs. In general, these

orbitals align with the three axes that exist in 3D space. Each of these orbitals

corresponds to a different value for m_l.

USE IT

EK 1.B.2
Student Edition
pp. 298–299

Draw a boundary surface diagram for each type of orbital for s and p orbitals.

EK 1.B.2
Student Edition p. 300

A subshell has the following values for its quantum numbers:
$n = 4, l = 1, m_l = -1, 0, 1$

What subshell do these quantum numbers identify?

_____

_____

_____

Section 7.8   Electron Configuration, *pp. 301–308*

Essential Knowledge Covered:
1.B.2

REVIEW IT

Using the word bank on the left, **fill in the blanks**. Words may be used more than once.

Word Bank

diamagnetic
electron configuration
Hund's rule
paramagnetic
Pauli exclusion
quantum numbers
shield

The four _____ allow us to identify each electron in a multi-electron atom. This gives us the _____ of the atom. This allows scientists to better understand the electronic behavior of each atom.

No two electrons have the same set of four _____. This is known as the _____ Principle. If the first three are the same for two electrons, each electron will have its own value of m_s value of either +1/2 or −1/2 to differentiate the two electrons. _____ substances contain net unpaired spins and are attracted by a magnetic. On the other hand, when there are no net unpaired spins, the substance is termed _____. As electrons build up in an atom, the electrons in lower energy levels are closer to the nucleus and tend to _____ the outer electrons from the nucleus. This effect tends to decrease the force of attracting acting on the outermost electrons.

As electrons continue to populate subshells, it has been found that the most stable arrangements of electrons in subshells occur when there is the greatest number of parallel spins. This is known as _____.

EK 1.B.2
Student Edition p. 303

What is the issue with representing the $1s^2$ as follows:

$\uparrow\uparrow$

_____

_____

_____

USE IT

EK 1.B.2
Student Edition p. 305

Draw electron arrangements for each electron configuration given below:

$1s^2\ 2s^2\ 2p^3$

What did you use to place the electrons in $2p$?

_____

_____

SUMMARIZE IT

EK 1.B.2

A student lists the electron arrangement for $4p^2$ as

Comment on the validity of this according to the rules of electron arrangements.

_____

_____

_____

EK 1.B.2

An atom of fluorine has 9 electrons. Write the four quantum numbers for each of the nine electrons in the ground state.

Electron	n	l	m_l	m_s	orbital
1					
2					
3					
4					
5					
6					
7					
8					
9					

Section 7.9 The Building-Up Principle, *pp. 308–313*

Essential Knowledge Covered:
1.B.2, 1.C.1

REVIEW IT

Using the word bank on the left, **fill in the blanks.**

Word Bank

actinides
Aufbau principle
lanthanides
noble gas
noble gas core
transition metals

The _____ states that as protons are added one by one to the nucleus of an atom to build up the elements, electrons are similarly added to the atomic orbitals. Electron configurations can be represented using a(n) _____, which shows the electron for the preceding _____ followed by the symbols for the electrons in the outermost orbitals.

_____ either have incompletely filled *d* subshells, or readily give rise to cations that have incompletely filled *d* subshells. _____ or rare earth metals have incompletely filled 4*f* shells, or readily give rise to cations that have incompletely filled 4*f* shells.

The elements in the last row are the _____ where most of these elements are not found in nature, but have been synthesized in a lab.

SUMMARIZE IT

EK 1.B.2, 1.C.1

Write the electron configuration for each element using the noble gas notation for each.

Al _____

N _____

Cu _____

_____

EK 1.B.2, 1.C.1

Identify the element that has the following ground state electron configurations.

$[Xe] 6s^2$ _____

$[Ar] 4s^2 3d^{10} 4p^3$ _____

$[Kr] 5s^1 4d^{10}$ _____

These questions were posed in the *Chemistry* chapter opener (page 274). Answer them using the knowledge you've gained from this chapter.

1. What is Planck's quantum theory?

   _____

   _____

   _____

2. What is the photoelectric effect?

   _____

   _____

   _____

3. What is the Bohr model of the hydrogen atom?

   _____

   _____

   _____

4. Can we determine the exact location of an electron bound to a nucleus? Why or why not?

   _____

   _____

   _____

5. What is an orbital? What is an electron configuration? What is shielding? What is the Aufbau principle?

   _____

   _____

   _____

As you work through your AP Focus Review Guide, keep this chapter's Big Ideas in mind:

BIG IDEA 1

The periodic table groups together elements that have similar structural features and display similar properties.

Section 8.1 Development of the Periodic Table, *pp. 327–329*

> ### Prerequisite Knowledge
>
> This section covers concepts you should know from other science courses. Review it to master the AP Essential Knowledge and prepare for the AP Exam.

Student Edition p. 327

REVIEW IT

What is the most important relationship between elements in the same group of the periodic table?

_____

_____

SUMMARIZE IT

Why was Newlands' Law of Octaves inadequate for elements beyond calcium?

_____

_____

_____

_____

Section 8.2   Periodic Classification of the Elements, *pp. 329–333*

Essential Knowledge Covered:
1.C.1

Word Bank

actinides
lanthanides
noble gases
representative
  elements
transition metals

REVIEW IT

For the following periodic table, list the type of element designated by letters A-H using the word bank on the left. Words may be used more than once.

																C	D
	A									B							
																	G
	F																

									H					
												E		

A.	B.	C.	D.
E.	F.	G.	H.

EK 1.C.1
Student Edition p. 330

Which letters in the periodic table above represent elements that would you expect to:

Have the same number of valence electrons?

_____

Have the same number of core electrons?

_____

Have similar physical properties?

_____

Section 8.2 Periodic Classification of the Elements (continued)

USE IT

EK 1.C.1
Student Edition p. 333

In the periodic table on page 331 of your textbook, which of the elements meets the following description.

A halogen that can form an ion with a 1-charge that is isoelectronic with Xe.

_____

A representative element that can form an ion with a 2+ charge that is isoelectronic with Ar.

_____

An element whose valence shell bears the following electron configuration: $3d^3$.

_____

SUMMARIZE IT

EK 1.C.1

The electron configurations of ions derived from representative elements follow a common pattern. What is that pattern, and how does it relate to the stability of these ions?

_____

_____

_____

Section 8.3 Periodic Variation in Physical Properties, *pp. 333–340*

Essential Knowledge Covered:
1.C.1

USE IT

EK 1.C.1
Student Edition p. 334

Complete the following Frayer Maps:

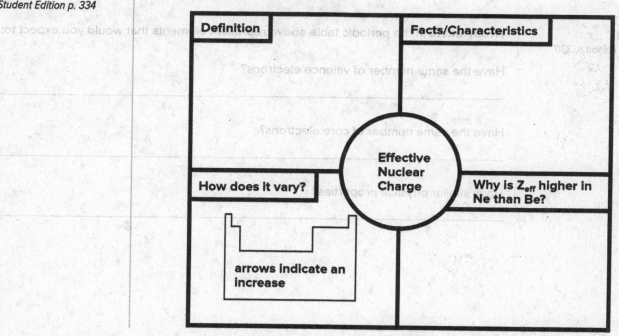

Definition	Facts/Characteristics

Atomic Radius

How does it vary?	Which atomic radius is higher, P or N? Why?
arrows indicate an increase	

SUMMARIZE IT

EK 1.C.1

How does the size of an atom change when it is converted from neutral to a cation? From neutral to an anion? Why?

_____

_____

_____

_____

Section 8.4 Ionization Energy, *pp. 340–345*

Essential Knowledge Covered:
1.C.1

EK 1.C.1
Student Edition pp. 342–344

REVIEW IT

Complete the Frayer map below:

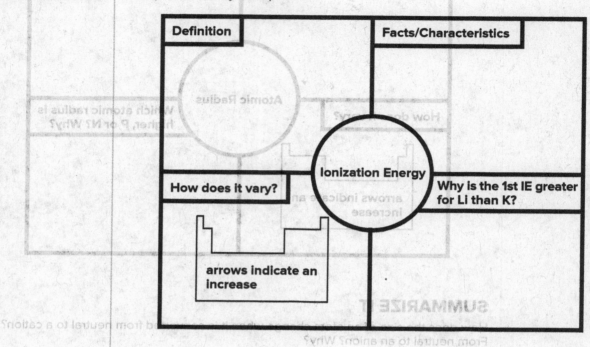

Definition	Facts/Characteristics
How does it vary?	Why is the 1st IE greater for Li than K?

Ionization Energy

arrows indicate an increase

USE IT

EK 1.C.1
Student Edition p. 342

Identify the following Period 3 element based on the given ionization energies (in kJ/mol).

IE1	IE2	IE3	IE4	IE5	IE6
1012	1903	2910	4956	6278	22230

_____
_____
_____
_____

EK 1.C.1
Student Edition p. 342

Identify the following Period 3 element based on the given ionization energies (in kJ/mol).

IE1	IE2	IE3	IE4	IE5	IE6
577	1816	2744	11576	14829	18375

_____
_____
_____
_____

Section 8.4   Ionization Energy (continued)

Section 8.5 Electron Affinity (continued)

EK 1.C.1

SUMMARIZE IT

Given the trend in ionization energy is an increase across a period, why does oxygen have a lower first ionization energy than nitrogen?

_____

_____

_____

_____

Section 8.5   Electron Affinity, *pp. 345–347*

Essential Knowledge Covered:
1.C.1, 1.D.1

REVIEW IT

EK 1.C.1
Student Edition p. 346

Rank the following compounds in order of increasing electron affinity:

Li, Na, K: _____

Li, C, F: _____

Br, I, Cl: _____

Sr, O, As: _____

USE IT

EK 1.C.1
Student Edition pp. 346–347

Explain the following trends in electron affinity (EA):

Noble Gases have negative EA values.

_____

_____

_____

Halogens have high EA values.

_____

_____

_____

_____

EA values increase across a period.

_____

_____

_____

EA values decrease down a group.

_____

_____

_____

Section 8.5   Electron Affinity (continued)

EK 1.C.1

SUMMARIZE IT

Why are there negative values for electron affinity but not for ionization energy?

_____

_____

_____

_____

EK 1.C.1

First electron affinities can be positive or negative, but second electron affinities (the energy change associated with adding a second electron) are always negative. Why?

_____

_____

_____

Section 8.6   Variation in Chemical Properties of the Representative Elements, *pp. 347–359*

Essential Knowledge Covered:
1.C.1

USE IT

EK 1.C.1
Student Edition p. 350

Many elements exhibit diagonal relationships, which are similarities between pairs of elements in different groups and periods of the periodic table, due to charge density similarity. If Li^+ and Mg^{2+} have a diagonal relationship, what must the approximate ratio of the cationic volumes be?

_____

_____

_____

EK 1.C.1
Student Edition p. 350

Water and certain other oxides possess a property known as amphotericity, which means it can act as both an acid and a base. Write the equation for the reaction of water with both a strong base (NaOH) and a strong acid (HCl):

_____

_____

_____

SUMMARIZE IT

EK 1.C.1

Until Bartlett synthesized $XePtF_6$, there were no known examples of compounds containing any of the noble gases. To this day no compounds containing Ne or He exist. Explain this apparent deviation from similarity in group behavior.

_____

_____

_____

_____

Section 8.6 Variation in Chemical Properties of the Representative Elements (continued)

EK 1.C.1

For each of the listed pair, give two examples of properties which indicate their chemical similarity:

Cl and Br

_____

_____

Cl and I

_____

_____

Li and K

_____

_____

Mg and Ca

_____

AP Reviewing the Essential Questions

These questions were posed in the *Chemistry* chapter opener (page 326). Answer them using the knowledge you've gained from this chapter.

1. How is the periodic table organized?

_____

_____

_____

2. What happens to the atomic radius of an atom when you add or remove an electron?

_____

_____

3. What is ionization energy, and how does it help demonstrate the shell model of electron structure?

_____

_____

_____

4. What is electron affinity? How does electron affinity vary with effective nuclear charge? Are non-metals more likely to gain or lose electrons? Are metals more likely to gain or lose electrons?

_____

_____

_____

_____

As you work through your **AP Focus Review Guide**, keep this chapter's **Big Ideas** in mind:

AP A LOOK AHEAD

BIG IDEA 1 Lewis structures of chemical compounds allow for a better understanding of how chemical bonds form between atoms.

BIG IDEA 2 The chemical and physical properites of compounds determine whether they will form ionic or covalent bonds.

BIG IDEA 3 The behavior of valence electrons can be described based on the electronegitivity differences between atoms in a bond.

BIG IDEA 5 The enthalpy of reactions can be understood through the energy associated with the formation and breakage of bonds.

Section 9.1 Lewis Dot Symbols, *pp. 369–370*

Essential Knowledge Covered:
1.C.1, 2.C.1, 2.C.2

EK 1.C.1, 2.C.1, 2.C.2
Student Edition p. 369

Word Bank

bond
configuration
four
isoelectronic
Lewis dot
noble gas
two
valence

REVIEW IT

Using the word bank on the left, **fill in the blanks.**

Gilbert Lewis was an American chemist who formulated an explanation for molecule and compound formations. He suggested that atoms will always combine to achieve a more stable electron _____. A maximum stability occurs when an atom is _____ with a _____. Given that it is the _____ electrons that cause an atom to not have the same electron _____ as a noble gas, a _____ symbol can be used to represent the electrons for atoms of an element. The symbol of the element is used, and is surrounded with one dot for each _____ electron associated with an atom of the element.

As an example, an element such as magnesium, located in group 2A will have _____ valence electrons, and so _____ dot(s) will be used around the symbol Mg to represent it. Once the number of electrons to be shown exceeds _____, the dots are paired up in a North-South-East-West arrangement around the symbol of the element.

These symbols can then be used to predict the type of _____ that will form between two atoms.

Section 9.1   Lewis Dot Symbols (continued)

SUMMARIZE IT

EK 1.C.1

Draw Lewis dot symbols for the following elements:

aluminum   **sulfur**

calcium   **argon**

Section 9.2   The Ionic Bond, *pp. 370–372*

Essential Knowledge Covered:
1.C.2, 2.C.2

SUMMARIZE IT

EK 1.C.2, 2.C.2

Use Lewis dot symbols and electron configuration to show the bonding that occurs when:

Aluminum and chlorine combine to form aluminum chloride.

Potassium and oxygen form potassium oxide

Essential Knowledge Covered:
1.B.1, 2.C.2

EK 1.B.1, 2.C.2
Student Edition
pp. 372–373

Word Bank

Born-Haber cycle
Coulomb's law
directly
electron affinity
inversely
ionization energy
ions
lattice energy
potential
stability

REVIEW IT

Using the word bank on the left, **fill in the blanks.**

Predicting which elements are likely to form ionic bonds requires an understanding of

the _____ of the ionic solid. This is dependent on all of the _____

and not simply a one anion, one cation interaction. A quantitative measure of the

stability is the _____. This is defined as the energy needed to completely

separate one mole of a solid ionic compound into gaseous ions.

    Since this cannot be measured directly, we must calculate it using _____.

This states that the _____ energy between two ions is

proportional to their charges and _____ proportional to the separation

distance of the charges.

    The _____ also can be used to determine the lattice energy.

This relates the lattice energy of an ionic compound to _____,

_____ and other atomic and molecular properties.

USE IT

EK 1.B.1, 2.C.2
Student Edition p. 373

Fill in the flow chart of the Born-Haber cycle for the reaction between sodium and
chlorine.

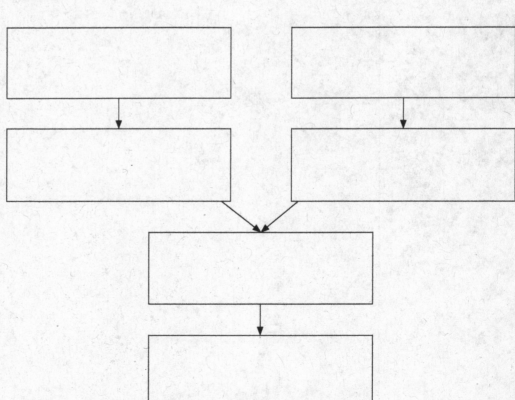

Section 9.3 Lattice Energy of Ionic Compounds (continued)

SUMMARIZE IT

EK 1.B.1, 2.C.2

Why are lattice energies always positive?

_____

_____

_____

EK 1.B.1, 2.C.2

Which physical property does lattice energy roughly correlate to? Explain your answer.

_____

_____

_____

_____

Section 9.4 The Covalent Bond, *pp. 377–380*

Essential Knowledge Covered:
2.C.1, 2.D.1, 2.D.2

REVIEW IT

Using the vocabulary on the left, **fill in the blanks.**

Word Bank

covalent bond
covalent compounds
double bond
hydrogen
Lewis structure
lone pairs
octet rule
single bond
triple bond

A _____ occurs when electrons are shared equally between two atoms. Compounds that only contain these types of bonds are referred to as _____ _____. The electrons in the bond are simultaneously attracted to the atoms being bonded together. Other electron pairs exist around some of the atoms; however these do not participate in the bonding between the atoms. These are referred to as _____.

The _____ of a covalent compound shows the bonding electron pair as either a pair of dots or a line, with the _____ surrounding the atoms to which they belong pictured around their symbols.

When atoms share valence electrons to form covalent bonds, they do so in order to be surrounded by 8 valence electrons (except for _____, which will share electrons until is it surrounded by just two electrons). This is referred to as the _____. This rule generally applies to elements in the second period of the periodic table, as they have subshells that can hold a total of eight electrons. In a _____, one pair of electrons hold the two elements in the bond together, while in a _____ or a _____, there are two or three electron pairs being shared.

SUMMARIZE IT

EK 2.C.1, 2.D.1, 2.D.2

Place the properties located to the left in the correct column below:

1. low melting point
2. strong intermo-
lecular forces
3. usually gases,
liquids or low
melting point
solids
4. solids at room
temperature
5. conduct electric-
ity in molten
state
6. high melting
point
7. soluble in water
8. conduct electric-
ity in aqueous
solution
9. relatively high
density
10. insoluble in
water
11. high boiling
point
12. poor electrical
conductivity

Covalent Compounds	Ionic Compounds

Section 9.5 Electronegativity, *pp. 380–383*

Essential Knowledge Covered:
1.C.1, 2.C.1

REVIEW IT

Using the vocabulary on the left, **fill in the blanks.** Words may be used more than once.

Word Bank

difference
electronegativity
ionic character
nonpolar
percent ionic character

_____ is a relative concept, in that the value itself does not indicate the behavior of the element, it is the _____ in electronegativity that determines the nature of the bond. When the _____ between the electronegativity values is greater than 2.0, the bond tends to be ionic and below 2.0, the bond tends to be covalent. Generally, when the electronegativity _____ is under 0.3, the bond is considered to be _____. Some chemists use _____ to describe the bonds. At a difference in electronegativity of 2.0, the bond is said to have a 50% _____. A difference in electronegativity of 0.0 would be 0% _____.

Section 9.5   Electronegativity (continued)

EK 1.C.1, 2.C.1
Student Edition p. 382

USE IT

Identify the type of bond that will form (ionic, polar covalent or covalent) between the two atoms in each of the following.

potassium and chlorine

calcium and oxygen

nitrogen and sulfur

carbon and iodine

SUMMARIZE IT

EK 1.C.1, 2.C.1

Draw a cartoon that helps explain the concept of the polar covalent bond between a fluorine and a hydrogen atom.

EK 2.C.1

Outline the similarity and difference between electronegativity and electron affinity.

_____

_____

_____

_____

Essential Knowledge Covered:
1.C.1, 2.C.1

EK 1.C.1, 2.C.1

SUMMARIZE IT

Create a flowchart to illustrate the steps needed to write Lewis structures of compounds.

EK 1.C.1, 2.C.1

Use this flowchart to create the Lewis structure for carbonic acid H_2CO_3. [Hint, the two hydrogen atoms are attached to two difference oxygen atoms and the carbon is surrounded by oxygen atoms]

Prerequisite Knowledge

This section covers concepts you should know from other science courses. Review it to master the AP Essential Knowledge and prepare for the AP Exam.

USE IT

Student Edition p. 387

Develop a flow chart diagram to explain how to assign a formal charge to an atom in a Lewis structure.

Student Edition p. 388

Use the flow chart created above to determine the formal charges for the nitrate ion, NO_3^-.

_____

Section 9.8   The Concept of Resonance, *pp. 390–392*

Essential Knowledge Covered:
2.C.4

REVIEW IT

Using the vocabulary on the left, **fill in the blanks.** Words can be used more than once.

Word Bank

actual
actual arrangement
bonding arrangement
Lewis
resonance structure

Many situations exist with _____ structures where more than one possible

_____ exists. Experimentally, it is found that the _____

_____ of the atoms does not correspond to any of these structures. When

this is the case, scientists use all possible _____ structures to represent the

molecule or complex ion. Each of these is referred to as a(n) _____.

This is defined as one of two or more _____ structures for a single molecule

or ion that cannot be represented using only one _____ structure. All of

these structures are separated with a double headed arrow to indicate that they are

_____.

   It must be remembered that none of these structures accurately represents the

_____ molecule. This concept is simply an attempt by science to explain

molecules using simple bonding models.

USE IT

EK 2.C.4
Student Edition p. 390

One resonance structure for the carbonate ion is given here. Create the
other possible resonance structure(s) for this ion.

SUMMARIZE IT

EK 2.C.4

Find the resonance structures of ozone on page 390 of your textbook. What scientific
evidence supports the fact that neither resonance structure for ozone actually exists?

_____

_____

_____

REVIEW IT

Using the vocabulary on the left, **fill in the blanks.**

Word Bank

aluminum
boron
coordinate covalent
halogens
octet
valence

Exceptions to the _____ rule exist in many situations. Elements in group 3A, particularly _____ and _____ tend to form compounds with fewer than eight electrons in the outer orbit. These elements form such compounds with _____. When these compounds participate in chemical reactions with compounds containing lone electron pairs around the central atom, often one of the lone electron pairs from one atom is used to form a bond. This type of bond is called a(n) _____ bond. This type of bond is the same as a normal covalent bond, it is just useful in helping to keep track of _____ electrons and assigning formal charge.

USE IT

EK 2.C.4
Student Edition
pp. 392–395

BCl₃
PCl₅
MgI₂
Li₂S
NO₂
SF₆
SO₄²⁻
NO

Exceptions to the octet rule generally fall into three categories: incomplete octet, odd number of electrons, and more than 8 valence electrons

With your understanding of Lewis structures, place each molecule or ion to the left in the appropriate column.

Incomplete Octet	Odd Number of Electrons	More than 8 Valence Electrons

Essential Knowledge Covered:
3.C.2, 5.C.1, 5.C.2

REVIEW IT

Using the vocabulary on the left, **fill in the blanks.** Words may be used more than once.

Word Bank

average
bond enthalpy
diatomic
difference
double
enthalpy
polyatomic
released
required
triple

There are several measures of the stability of a molecule, one such measure is

_____. This is the _____ change required to break a particular

bond in one mole of gaseous molecules. For more stable bonds, more _____

is needed to break the bond. These values can also be directly measured for

_____ molecules containing unlike elements such as HF, or for molecules

containing _____ or _____ bonds such as oxygen gas or nitrogen

gas. In _____ molecules, the process is much more complicated, so instead,

scientists use _____ bond enthalpies for an approximate calculation.

   The approximate _____ of a reaction can be calculated by remembering

that energy is _____ to break bonds while energy is _____ when

bonds are formed. The _____ between the total _____ of the

reactants and the total _____ of the products will give an approximate

_____ of the reaction.

USE IT

EK 3.C.2, 5.C.2
Student Edition p. 399

What does the segment from a Bond Enthalpy chart illustrate with respect to the enthalpy of a bond and the type of bond?

Bond	Enthalpy (kJ/mol)
C—N	276
C=N	615
C≡N	891

_____

_____

SUMMARIZE IT

EK 5.C.1, 5.C.2

Below is a section of a chart on bond enthalpies:

Bond	Bond Enthalpy (kJ/mol)	Bond	Bond Enthalpy (kJ/mol)
H—H	436.4	C—I	240
H—N	393	C—P	263
H—O	460	C—S	255
H—S	368	C≡S	477
H—P	326	N—N	193
H—F	568.2	N≡N	418
H—Cl	431.9	N≡N	941.4
H—Br	366.1	N—O	176
H—I	298.3	N≡O	607
C—H	414	O—O	142
C—C	347	O≡O	498.7
C≡C	620	O—P	502
C≡C	812	O≡S	469
C—N	276	P—P	197
C≡N	615	P≡P	489
C≡N	891	S—S	268
C—O	351	S≡S	352
C═O†	745	F—F	156.9

Use this chart to determine the approximate enthalpy of reaction for

$$C_2H_2 + O_2 \rightarrow CO_2 + H_2O$$

These questions were posed in the *Chemistry* chapter opener (page 368). Answer them using the knowledge you've gained from this chapter.

1. How are electron configurations used to determine Lewis dot symbols?

2. What is an ionic bond? What are some characteristics of ionic compounds?

3. How does ionic charge and ion separation influence potential energy?

4. What is a covalent bond? What are some characteristics of covalent compounds?

5. What is electronegativity? How does electronegativity vary within the periodic table? What is a polar bond?

6. What is bond enthalpy? How are bond enthalpies used to determine the enthalpy of reaction?

CHEMICAL BONDING II: MOLECULAR GEOMETRY AND HYBRIDIZATION OF ATOMIC ORBITALS

As you work through your AP Focus Review Guide, keep this chapter's Big Ideas in mind:

AP ► A LOOK AHEAD

BIG IDEA 2 The three-dimensional array of elements in a compound plays a substatial role in determining the chemical and physical properties of that compound.

Section 10.1 Molecular Geometry, *pp. 413–423*

Essential Knowledge Covered:
2.C.4

Word Bank

Bent
Linear
Octahedral
Seesaw
Square planar
Square pyramidal
T-shaped
Tetrahedral
Trigonal bipyramidal
Trigonal planar

REVIEW IT

Complete the following table of electron pair arrangements and their corresponding molecular geometries, using the word bank on the left.

Total Electron Pairs	Bonding Pairs	Lone Pairs	Arrangement of Electron Pairs	Molecular Geometry
2	2	0		
3	3	0		
4	4	0		
5	5	0		
6	6	0		
3	2	1		
4	3	1		
4	2	2		
5	4	1		
5	3	2		
5	2	3		
6	5	1		
6	4	2		

USE IT

EK 2.C.4
Student Edition p. 413

In the table on page 102, what trend exists between electron pair arrangement and total electron pairs? How does VSEPR theory explain this trend?

_____

_____

_____

_____

EK 2.C.4
Student Edition p. 413

What molecular geometry will methane (CH_4) adopt? Use VSEPR theory to explain this geometry and draw a picture that supports your reasoning.

SUMMARIZE IT

EK 2.C.4

Which of the following two structures (A or B) indicates the geometry that SF_4 adopts? What is that geometry, and why does it adopt your chosen structure instead of the other?

A   B

_____

_____

_____

_____

Essential Knowledge Covered:
2.C.1

EK 2.C.1
Student Edition p. 423

USE IT

Classify the following compounds as polar or nonpolar.

CH_2Cl_2 _____

H_2O _____

C_2H_4 _____

$AlCl_3$ _____

HF _____

EK 2.C.1
Student Edition p. 423

Add the names of the compounds in the table to the appropriate section of the Venn Diagram.

Cl≡Cl	H≡Cl	H_2O
BeH_2	CO_2	CO
HF	H_2SO_4	SF_4

Compounds with Polar Bonds   **Polar compounds**

Section 10.2 Dipole Moments (continued)

SUMMARIZE IT

EK 2.C.1

How is it possible to have a compound that contains polar bonds but is nonpolar?

_____

_____

EK 2.C.1

Is it possible to have a compound that is polar but does not contain polar bonds?

_____

EK 2.C.1

Why do the dipole moments of the hydrohalic acids decrease from HF to HI?

_____

_____

_____

Section 10.3 The Valence Bond Theory, *pp. 429–431*

Essential Knowledge Covered:
2.C.4

USE IT

EK 2.C.4
Student Edition p. 433

Turn to Figure 10.5 on page 430 of your textbook, which shows the potential energy of the system as two atoms approach one another. Use this graph to answer the questions that follow.

Why is potential energy zero at point the point of separation?

_____

_____

Where is the potential energy at its minimum?

_____

_____

Where is the system most stable?

_____

SUMMARIZE IT

EK 2.C.4

Why is Valence Bond theory considered an improvement over Lewis or VSEPR theory?

_____

_____

_____

_____

EK 2.C.4

The law of conservation of energy dictates that the total amount of energy in the universe is constant. If two atoms that approach one another decrease in potential energy, where does that energy go?

_____

_____

Section 10.4 Hybridization of Atomic Orbitals, *pp. 431–440*

Essential Knowledge Covered:
2.C.4

AP Focus note: sp^{3}d and sp^{3}d^2 hybridization is not a part of the AP Curriculum, but other properties of these molecules may be on the exam.

Word Bank

atomic orbitals
conserved
covalent bonds
different
hybridization
molecular geometries
overlap
same

REVIEW IT

Using the word bank on the left, **fill in the blanks.**

_____ is defined as the mixing of _____ in a central atom to generate a set of hybrid orbitals. Valence Bond theory relies on hybridized orbitals to accurately describe and explain _____, though it is important to note that hybridization is a theoretical construct. Hybridization can be further restricted to the mixing of at least two _____ atomic orbitals, such as *s* and *p*-type orbitals. Orbitals can be considered to be _____, as the number of orbitals that are being hybridized is the _____ as the number of hybrid orbitals.

_____ arise from the overlap of hybrid orbitals with either hybrid or non-hybrid orbitals in polyatomic molecules; in both cases the electron pair is defined as residing within the _____ area.

EK 2.C.4
Student Edition p. 440

USE IT

What is the hybridization state of Xe in XeF_4? Show your answer with a drawing the appropriate orbital diagram.

SUMMARIZE IT

EK 2.C.4

Why must hybridization occur between different orbital subtypes?

_____

_____

_____

EK 2.C.4

Why won't an isolated atom form hybrid orbitals?

_____

_____

_____

EK 2.C.4

Describe the state of hybridization of the indicated atom in the following molecules:

P in PCl_5:   Se in SeF_6:

C in CO: C in CO_2:

Section 10.5 Hybridization in Molecules Containing Double and Triple Bonds, *pp. 440–443*

Essential Knowledge Covered:
2.C.4

REVIEW IT

Complete the following comparison charts using the word bank to the left:

Word Bank

hybridized
pi bond
restricted
sigma bond
unhybridized
unrestricted

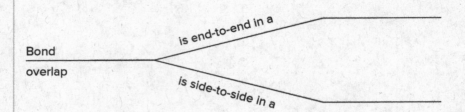

Bond overlap — is end-to-end in a ___ / is side-to-side in a ___

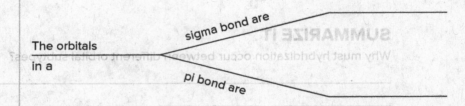

The orbitals in a — sigma bond are ___ / pi bond are ___

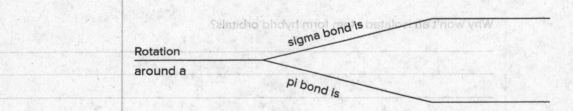

Rotation around a — sigma bond is ___ / pi bond is ___

USE IT

EK 2.C.4
Student Edition p. 441

Define the general rule that exists between hybridization state and whether a double, or triple bond forms.

_____

_____

SUMMARIZE IT

EK 2.C.4

Compounds A and B below are identical compounds, but compounds C and D are not. Explain why, using a discussion of sigma and pi bonds:

_____

_____

_____

_____

_____

Section 10.6   Molecular Orbital Theory, *pp. 443–446*

> ### Extending Knowledge
>
> This section takes the AP Essential Knowledge you have learned further, and may provide illustrative examples useful for the AP Exam.

SUMMARIZE IT

Why are molecular orbitals described as bonding or antibonding?

_____

_____

_____

What similarities do Molecular Orbital and Valence Bond theory have in their treatment of the bond types in multiple (double or triple) bond cases?

_____

_____

Section 10.7   Molecular Orbital Configurations, *pp. 446–452*

> ### Extending Knowledge
>
> This section takes the AP Essential Knowledge you have learned further, and may provide illustrative examples useful for the AP Exam.

REVIEW IT

Word Bank

antibonding
atomic orbitals
bonding
high
Hund's Rule
low
molecular orbitals
number of electrons
opposite
two
zero

Using the word bank on the left, **fill in the blanks** in the following discussion of the rules governing molecular orbital configurations.

1. The number of molecular orbitals formed is always the same as the number of _____ that combined.

2. The more stable the _____ molecular orbital, the less stable the corresponding _____ molecular orbital.

3. The filling of molecular orbitals proceeds from _____ energy to _____. In a stable molecule, the _____ in bonding orbitals is always greater than the number of electrons in antibonding orbitals.

4. Each molecular orbital can accommodate up to _____ electrons with _____ spins.

5. When electrons are added to molecular orbitals of the same energy, the most stable arrangement is given by _____.

6. The number of electrons in the _____ is the same as the number of the electrons in the bonding atoms.

7. Bond order predicts a compound's likelihood to exist; bond orders greater than _____ are indicative of stable compounds.

Section 10.8   Delocalized Molecular Orbitals, *pp. 452–455*

Essential Knowledge Covered:
2.C.4

EK 2.C.4
Student Edition p. 453

USE IT

Benzene is commonly drawn in the following representation:

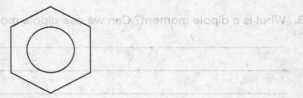

Describe how this structure is representative of the true structural nature of benzene.

_____

_____

_____

_____

_____

_____

_____

SUMMARIZE IT

EK 2.C.4

Explain, using the concept of delocalized molecular orbitals, why compound A is more stable than compound B:

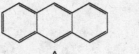

A   B

_____

_____

_____

_____

_____

_____

These questions were posed in the *Chemistry* chapter opener (page 412). Answer them using the knowledge you've gained from this chapter.

1. What is the VSEPR model? How is the VSEPR model used to predict molecular geometry?

_____

_____

_____

2. What is a polar bond?

_____

_____

3. What is a dipole moment? Can we use dipole moments to confirm molecular geometry?

_____

_____

_____

4. What is the valence bond theory?

_____

_____

_____

_____

5. What is molecular orbital theory?

_____

_____

_____

_____

6. What are delocalized orbitals?

_____

_____

_____

_____

_____

_____

INTERMOLECULAR FORCES AND LIQUIDS AND SOLIDS

As you work through your AP Focus Review Guide, keep this chapter's Big Ideas in mind:

AP ▸ A LOOK AHEAD

BIG IDEA 2 Many of the physical properties of matter arise from the way neighboring molecules interact.

BIG IDEA 5 Phase changes either require heat or liberate heat based on how the intermolecular forces change.

BIG IDEA 6 The phases of matter exist in equilibrium with one another, and this equilibrium can be exploited.

Section 11.1   The Kinetic Molecular Theory of Liquids and Solids, *pp. 466–467*

Essential Knowledge Covered:
2.A.1

SUMMARIZE IT

EK 2.A.1

Molecules in motion have kinetic energy. Molecules also have potential energy by virtue of their attraction to one another. Describe how these two types of energy exist in each phase of matter.

Solid:
_____
_____
_____

Liquid:
_____
_____
_____

Gas:
_____
_____

REVIEW IT

Vocabulary:

dipole-dipole forces
dipole-induced dipole
  forces
dispersion forces
hydrogen bond
ion-dipole forces
intermolecular forces
van der Waals forces

Complete the following concept web using the vocabulary on the left.

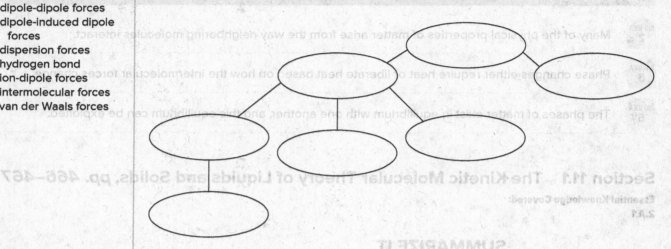

EK 2.B.2
Student Edition p. 472

In the following table, complete the columns that identify the partners in each type of intermolecular force.

In a:	a(n):	is attracted to a(n):
dipole-dipole force		
dipole-induced dipole force		
dispersion force		
hydrogen bond		
ion-dipole force		
ion-induced dipole force		

USE IT

EK 2.B.2
Student Edition p. 472

Indicate (Y/N) if the following compounds that can form hydrogen bonds with water.

Compound	Hydrogen bond with water?
NH_3	
CH_4	
CH_3CO_2H	
HF	
HBr	

Section 11.2   Intermolecular Forces (continued)

EK 5.D.1
Student Edition p. 471

What type(s) of intermolecular force(s) is/are present between the following pairs of compounds?

Force Type(s)	Compound 1	Compound 2
	Water	Na^+
	CH_4	CH_4
	F^-	CH_4
	CH_3OCH_3	H_2O

SUMMARIZE IT

EK 2.B.2

What are the criteria that allow two molecules to interact via a hydrogen bond? Why is the hydrogen bond considered a special type of dipole-dipole interaction?

_____

_____

_____

_____

EK 2.B.1

Why are dispersion forces always present between neighboring molecules?

_____

_____

_____

_____

Section 11.3   Properties of Liquids, *pp. 473–477*

Essential Knowledge Covered:
2.A.1

USE IT

EK 2.A.1
Student Edition p. 474

Using the structure below, explain why water flows faster than glycerol.

_____

_____

EK 2.A.1
Student Edition p. 476

Explain why viscosity tends to decrease as temperature increases.

_____

_____

_____

EK 2.A.1

SUMMARIZE IT

Photosynthesis in trees occurs in the leaves, far from the ground where the water that is also required is obtained. The trunk of a tree has a certain type of pore that approximates a capillary tube within its structure and which runs the length of the trunk. How do the properties of water allow it to travel the length of a tree to where it is needed?

_____

_____

_____

Section 11.4   Crystal Structure, *pp. 477–483*

Extending Knowledge

This section takes the AP Essential Knowledge you have learned further, and may provide illustrative examples useful for the AP Exam.

REVIEW IT

Word Bank

closest packing
coordination number
crystalline solid
hexagonal close-
    packed
intermolecular
    interactions
rigid
simple cubic
unit cell
well-defined
    arrangement

Using the word bank on the left, **fill in the blanks.**

A _____ is any solid that has _____ and _____

of its component ions, atoms, or molecules; that is, the components occupy very specific

positions in the solid lattice. The basic repeating structural unit is the _____, which

can take on seven different fundamental forms based on how the component particles

fit together; the aim being the maximization of _____. Unit

cells are quantified based on their _____, or the number of compo-

nent particles that surrounds a central particle in a lattice. Many of the packing types,

such as _____, are somewhat inefficient, as there is a good deal of empty

space in the lattice. _____, however, is the method by which the most effi-

cient arrangement of spheres is assembled and is seen in a few types of structures,

such as _____ (hcp) or cubic close-packed (ccp).

Section 11.5   X-Ray Diffraction by Crystals, *pp. 483–486*

> ## Extending Knowledge
> This section takes the AP Essential Knowledge you have learned further, and may provide illustrative examples useful for the AP Exam.

REVIEW IT

Student Edition p. 483

What is X-ray diffraction? Why is it used?

_____

_____

_____

Student Edition p. 485

Why is X-ray diffraction only useful for substances whose structures are ordered crystals?

_____

_____

_____

Section 11.6   Types of Crystals, *pp. 486–492*

Essential Knowledge Covered:
2.C.3, 2.D.1, 2.D.2, 2.D.3

REVIEW IT

EK 2.C.3, 2.D.2
Student Edition p. 491

Complete the table summarizing the properties of the four main types of crystals.

Crystal type:	Ionic	Covalent	Molecular	Metallic
Particle type				
Force that holds adjacent crystals together				
Atom type present				
Hardness				
Thermal Conductivity				
Electrical conductivity				
Example				

Section 11.6　Types of Crystals (continued)

USE IT

EK 2.C.3, 2.D.2
Student Edition p. 491

What aspect of metallic crystal structure makes metals such good conductors of electricity?

_____

_____

_____

EK 2.D.1
Student Edition p. 487

Why are ionic compounds poor conductors, while ionic solutions are good conductors?

_____

_____

_____

SUMMARIZE IT

EK 2.D.3

Why do graphite and diamond, which are both allotropes of carbon, have such wildly different hardness properties?

_____

_____

_____

_____

Section 11.7　Amorphous Solids, *pp. 492–493*

Extending Knowledge

This section takes the AP Essential Knowledge you have learned further, and may provide illustrative examples useful for the AP Exam.

SUMMARIZE IT

Why do amorphous solids form? What is one method by which an amorphous solid could be formed in the laboratory?

_____

_____

_____

Quartz glass is colorless, but other types of glass can be made to have virtually any color that is desired. How is this done?

_____

_____

Section 11.8 Phase Changes, *pp. 493–503*

Essential Knowledge Covered:
2.A.2, 5.B.3, 5.D.1, 6.A.1

Vocabulary:

boiling point
condensation
critical pressure
critical temperature
deposition
evaporation
freezing point
melting point
phase changes
sublimation
vaporization

REVIEW IT

Using the vocabulary on the left, **fill in the blanks:**

Transitions between solid, liquid, and gas are referred to globally as _____.

When a liquid is converted to a gas, it is called _____ or _____.

The gas phase molecules re-entering the liquid phase is known as _____.

There exists a temperature above which no condensation is possible; this is the

_____, and the pressure necessary to condense a substance at

its critical temperature is its _____. In the vaporization process, the

point at which the vapor pressure is equal to the external pressure is known as the

_____ of the liquid. To convert a liquid to a solid, the substance must

undergo a process called freezing, which occurs at the substance's _____.

The reverse process, melting, occurs at the substance's _____. Both of these

points are temperatures at which point the solid and liquid phases coexist in some

equilibrium. A solid can be converted directly to a gas (bypassing the liquid phase) via

_____, and a gas can solidify directly by _____.

USE IT

EK 5.B.3, 6.A.1
Student Edition
pp. 494–502

Identify the processes that interconvert between the three phases of matter as shown below:

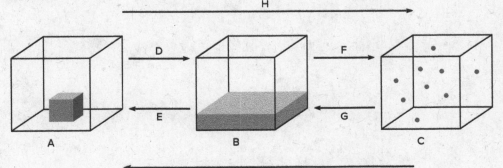

Arrow	Process
D	
E	
F	
G	
H	
I	

EK 2.A.2
Student Edition p. 496

There exists a quantitative relationship between vapor pressure and temperature, which is described by the Clausius-Clapeyron equation. Why does increasing temperature cause vapor pressure to increase?

_____

_____

SUMMARIZE IT

EK 2.A.2

If the vapor pressure of ethanol is 115 mm Hg at 34.9°C, what is the temperature at which the vapor pressure is 760 mm Hg if ΔH_{vap} is 38.6 KJ/mol?

EK 5.D.1

Why is condensation not possible if a substance's temperature rises above its critical point? Explain, from a kinetic molecular perspective.

_____

_____

_____

Extending Knowledge

This section takes the AP Essential Knowledge you have learned further, and may provide illustrative examples useful for the AP Exam.

USE IT

Use the phase diagram to answer the questions that follow.

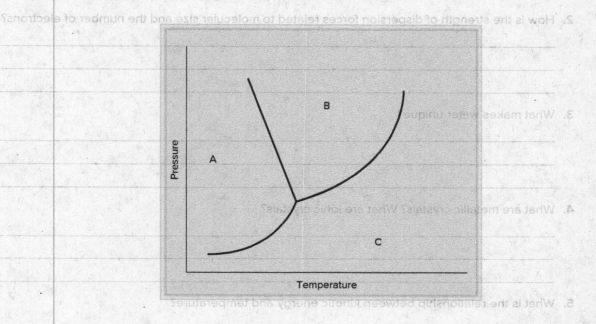

Student Edition
pp. 503–504

What would happen if:

1. we began at point A, and raised the temperature at constant pressure

_____

_____

_____

2. we began at point C and reduce the temperature at constant pressure

_____

_____

_____

3. we began at point B and reduce the pressure at constant temperature

_____

_____

_____

AP Reviewing the Essential Questions

These questions were posed in the *Chemistry* chapter opener (page 465). Answer them using the knowledge you've gained from this chapter.

1. How are the properties of liquids and solids influenced by interactive forces?

_____

_____

_____

2. How is the strength of dispersion forces related to molecular size and the number of electrons?

_____

_____

_____

3. What makes water unique?

_____

_____

_____

_____

4. What are metallic crystals? What are ionic crystals?

_____

_____

_____

_____

5. What is the relationship between kinetic energy and temperature?

_____

_____

_____

6. Are phase transitions chemical or physical changes?

_____

_____

_____

7. Are phase changes reversible?

_____

_____

_____

_____

PHYSICAL PROPERTIES OF SOLUTIONS

As you work through your AP Focus Review Guide, keep this chapter's Big Ideas in mind:

AP > A LOOK AHEAD

BIG IDEA 2 The structure of solutes and solvents, as well as their intermolecular forces, determines the properties of a solution.

BIG IDEA 5 Entropy and enthalpy of systems play a large influence the properties of solutions.

Section 12.1   Types of Solutions, *pp. 519–520*

Essential Knowledge Covered:
2.A.3

EK 2.A.3
Student Edition
pp. 519–520

Word Bank

capacity
crystallization
homogeneous
saturated
solute
solvent
supersaturated
unsaturated

REVIEW IT

Using the word bank on the left, **fill in the blanks.** Words may be used more than once.

A solution is a _____ mixture of two or more substances. Here we will focus on solutions where at least one of the substances is a liquid. This liquid or the liquid of the largest volume if there are two is called the _____ and the substance of the lesser quantity is called the _____.

Chemists characterize solutions by their _____ to dissolve a _____. A(n) _____ solution contains the maximum amount of solute that will dissolve in a solvent at a specific temperature. A(n) _____ solution contains less than this maximum amount, and a _____ solution contains more than this maximum amount. _____ solutions are unstable and can have a solid form in a process called _____ by either agitation or seeding of the solution.

EK 2.A.3
Student Edition p. 520

How do scientists usually differentiate the solid formation in a precipitation reaction from the solid formation in a crystallization process?

_____

_____

_____

Essential Knowledge Covered:
2.A.3, 2.B.3, 5.E.1

REVIEW IT

EK 2.A.3, 2.B.3, 5.E.1
Student Edition p. 520

What are the three interactions that must be considered with respect to the ease at which solute particles replace solvent particles in a solution?

_____

_____

_____

EK 2.A.3, 2.B.3, 5.E.1
Student Edition p. 520

What two factors govern all processes in the universe that are involved in the process of the formation of a solution?

_____

_____

USE IT

EK 2.A.3, 2.B.3, 5.E.1
Student Edition p. 520

Create a schematic diagram to outline the solution process at the molecular level.

SUMMARIZE IT

EK 2.B.3

In which liquid would sodium chloride be more soluble, water or CCl_4? Explain your answer.

_____

_____

_____

_____

Section 12.3   Concentration Units, *pp. 522–526*

Essential Knowledge Covered:
2.A.3

EK 2.A.3
Student Edition
pp. 522–523

Word Bank

divided
molality
molarity
mole fraction
percent by mass
ratio
unitless

REVIEW IT

Using the word bank on the left, **fill in the blanks.** Words may be used more than once.

The _____ is the _____ of the mass of solute to the mass of solution, multiplied by 100%. The value itself is _____. The _____ is a measure of concentration that involves the moles of component A in the solution _____ by the sum of the moles of all components. It is given the symbol *X*. This value is _____.

The _____ is a concentration measure that is the ratio of the number of moles of solute to the number of liters of solution. _____ is a concentration measure that is the ratio of the number of moles of the solute to the mass of the solvent, in kg.

EK 2.A.3
Student Edition
pp. 522–523

Write the equation for the following concentration expressions: percent by mass, mole fraction, molarity.

USE IT

EK 2.A.3
Student Edition p. 523

A sample of 0.500 g of sodium chloride is dissolves in 49.50 g of water. What is the percent by mass of this solution?

SUMMARIZE IT

EK 2.A.3

Place each measure of the concentration of a solution in the correct column.

Word Bank

molality
molarity
mole fraction
percent by mass

Has Units	Unitless
Temperature Dependent	**Independent of Temperature**

Section 12.4   The Effect of Temperature on Solubility, *pp. 527–529*

Extending Knowledge

This section takes the AP Essential Knowledge you have learned further, and may provide illustrative examples useful for the AP Exam.

REVIEW IT

Student Edition p. 528

Turn to Figure 12.3 on page 527 of your textbook. Rank the solids in terms of their solubility at 80°C, from lowest to highest.

_____

_____

SUMMARIZE IT

In which situation(s) is fractional crystallization most efficient at separating impurities in a solution?

_____

_____

REVIEW IT

Student Edition p. 530

Use Henry's law to explain why a soda bottle that has been shaken spills when opened.

_____

_____

SUMMARIZE IT

The bends is a condition that scuba divers must be aware of as they surface from a deep dive. The pressurized air that they have been breathing while diving contains nitrogen gas that, at depths under the water, will dissolve into the blood and surrounding tissues. What will happen to this dissolved gas if the diver surfaces too quickly?

_____

_____

_____

_____

Prerequisite Knowledge

This section covers concepts you should know from other science courses. Review it to master the AP Essential Knowledge and prepare for the AP Exam.

REVIEW IT

Word Bank

colligative
dilute
fractional distillation
mole fraction
nonelectrolyte
nonvolatile
osmotic
partial
Raoult's
solute
sum
volatile

Using the word bank on the left, **fill in the blanks.** Words may be used more than once.

A(n) _____ property is one that only depends on the number of _____ particles in a solution, not on their nature or behavior in the solution. In this section, the solution will be a(n) _____ solution and for the purposes of the properties discussed, the solution must be relatively _____.

If a solute is _____ meaning that it does not have a measureable vapor pressure, the vapor pressure of the solution will always be less than that of the pure solvent. This is a statement of _____ law. In this law, the _____ is used to find the vapor pressure of the solvent.

If both components of the solution are _____, the vapor pressure of the solution is the _____ of the individual _____ pressures.

_____ is a procedure for separating liquid components of a solution based on differing boiling points. The temperature is held just above the boiling point of the more _____ liquid, and the vapors are condensed. This solution still contains a small quantity of the less _____ liquid, so often the process needs to be repeated to increase the purity of the distillate.

All processes involving freezing, boiling or _____ pressure of a solution all are affected by the interactions of particles in the solution.

Student Edition p. 533

USE IT

The water vapor pressure at 40°C is 55.32 mm Hg. Use the Raoult's law equation $P_1 = X_1 P_1°$ to determine the vapor solution of a solution made by dissolving 153.2 g of a nonvolatile solid with a molar mass of 134.1 g/mol in 575 mL of water at 40°C. Here, assume that the density of the solvent is 1.00 g/mL.

SUMMARIZE IT

Explain why the vapor pressure of a solution is always less than that of the pure solvent.

_____

_____

_____

_____

Prerequisite Knowledge

This section covers concepts you should know from other science courses. Review it to master the AP Essential Knowledge and prepare for the AP Exam.

REVIEW IT

Using the word bank on the left, **fill in the blanks.**

Word Bank

dissociate
electrolyte
electrostatic
ions
ion pair
larger
smaller
van't Hoff factor

When dealing with a(n) _____ solution, the fact that the solution will _____ into _____ must be taken into account. The reason that this must be taken into account is that the total number of particles determines colligative properties, and each electrolyte particle will _____ into two particles. This is taken into account using the _____.

Colligative properties of electrolyte solutions are usually _____ than anticipated due to the fact that at higher concentrations, _____ forces must be taken into account. A(n) _____ is made up of one or more cations and one or more anions held together by these forces. The presence of these ions reduces the number of particles available in solution, resulting in a reduction in the colligative properties.

Student Edition p. 545

Beside each material listed below indicate if you expect the measured value for *i* to be lower than calculated, the same as calculated, or higher than calculated for a constant temperature and concentration of solution.

Solute	*i* calculated	Expected *i* measured
$MgCl_2$	2.9	
KBr	1.8	
Glucose	1.1	
FeI_2	3.1	

> # Extending Knowledge
>
> This section takes the AP Essential Knowledge you have learned further, and may provide illustrative examples useful for the AP Exam.

SUMMARIZE IT

Illustrate how hydrophobic colloids are stabilized. Your drawing should include colloidal particles, ions, and indications of any forces at work.

Create a diagram illustrating how soap can remove grease from your hands.

Reviewing the Essential Questions

These questions were posed in the *Chemistry* chapter opener (page 518). Answer them using the knowledge you've gained from this chapter.

1. What are some types of solutions? What types of interactions occur in the solution process? What is the heat of solution? What is solvation?

_____

_____

_____

_____

2. What are miscible solutions?

_____

_____

3. What factors besides energy govern the solution process?

_____

_____

_____

4. What is molarity? What is fractional distillation? What occurs when ionic compounds dissolve in water?

_____

_____

_____

_____

As you work through your AP Focus Review Guide, keep this chapter's Big Ideas in mind:

AP   A LOOK AHEAD

BIG IDEA 4 Chemical kinetics is concerned with the many factors, from the particle to system level, that influence the rate at which chemical reactions occur.

Section 13.1   The Rate of a Reaction, *pp. 563–570*

Essential Knowledge Covered:
4.A.1, 4.A.3

EK 4.A.1, 4.A.3
Student Edition
pp. 563–565

Word Bank

average
decrease
higher
increase
instantaneous
kinetics
lower
proportional
rate
stoichiometry
yield

REVIEW IT

Using the word bank on the left, **fill in the blanks.** Some words may be used more than once, or not at all.

Chemical _____ in chemistry is the study of the rate at which a chemical

reaction occurs. The reaction _____ is a measure of the change in concen-

tration of a reactant or product over time. While it might make sense to look at increas-

ing the _____ of a reaction, scientists find it more important to increase the

_____ of the reaction instead.

At the start of a reaction, the _____ rate of the reaction will be _____

than towards the completion of the reaction. This is simply due to the _____

in the number of reactant particle collisions that will cause the formation of product.

This tells us that, in general, the rate of a reaction is _____ to the concentra-

tion of reactants. It also makes sense that the rate of a chemical reaction can be

dependent on the _____ of the reaction.

USE IT

EK 4.A.1
Student Edition p. 566

In a reaction involving two or more reactants, how do scientists determine the effect on
the reaction rate by changes to just one of the reactants?

_____

_____

_____

EK 4.A.1
Student Edition p. 564

For each of the reactions given, determine the factor that can be monitored to measure the rate of each reaction. Include an explanation of how this quantity will change as the reaction proceeds.

$H_2O_2(l) \rightarrow H_2O(l) + O_2(g)$

_____

_____

$NH_4Cl(s) \rightarrow NH_4^+(aq) + Cl^-(aq)$

_____

_____

$Pb(NO_3)_2(aq) + KI(aq) \rightarrow PbI_2(s) + KNO_3(aq)$

_____

_____

SUMMARIZE IT

EK 4.A.1, 4.A.3

The table below gives information regarding the rate of reaction of reactants A and B. For the data given, reactant A was in a low concentration and reactant B is a very high concentration. Use the information in the table to fill in the last column for the rate constant.

Time (seconds)	[A] (M)	Rate (M/s)	$k = \dfrac{rate}{[A]}$ (s^{-1})
0.0	0.0184	5.75×10^{-5}	
10.0	0.0160	5.00×10^{-5}	
20.0	0.0139	4.35×10^{-5}	
30.0	0.0121	3.79×10^{-5}	
40.0	0.0105	3.29×10^{-5}	
50.0	0.00917	2.87×10^{-5}	
60.0	0.00797	2.49×10^{-5}	

Section 13.2   The Rate Law, *pp. 571–574*

Essential Knowledge Covered:
4.A.1, 4.A.2, 4.A.3

EK 4.A.2
Student Edition
pp. 571–572

REVIEW IT

What is the overall *reaction order* of a chemical reaction and how is it found?

_____

_____

EK 4.A.1, 4.A.2
Student Edition p. 571

USE IT

What must be true of the exponents of the rate law for the reaction $2SO_2(g) + O_2(g) \rightarrow 2SO_3(g)$ when the rate law is expressed as $rate = k[SO_2(g)]^x[O_2(g)]^y$?

_____

_____

EK 4.A.2, 4.A.3

SUMMARIZE IT

Complete the following summary of the points of the rate law:

1. Rate laws are always determined experimentally.

2. The reaction order is defined in terms of _____.

3. The overall reaction order is found by _____

   _____.

4. The order of a reactant is not related to _____

   _____.

EK 4.A.2

For the data given, determine the rate law expression for the reaction
$2NO(g) + 2H_2(g) \rightarrow N_2(g) + 2H_2O(g)$

Trial	[NO]	[H$_2$]	Initial rate
1	0.001	0.004	0.004
2	0.002	0.004	0.016
3	0.003	0.004	0.036
4	0.004	0.001	0.016
5	0.004	0.002	0.032
6	0.004	0.003	0.048

Section 13.3 The Relation Between Reactant Concentration and Time, *pp. 575–587*

Essential Knowledge Covered:
4.A.1, 4.A.2

EK 4.A.2
Student Edition p. 575

REVIEW IT

In the expression $\ln[A]_t = -kt + \ln[A]_0$, explain the type of relation this represents and describe the graph that would be created if this relation were to be graphed.

_____

_____

_____

EK 4.A.1, 4.A.2
Student Edition
pp. 576–577

USE IT

In a first order reaction of A → B, it is found that the rate constant is 8.25×10^{-4} s^{-1} at 200°C. The initial concentration of A is 0.45 M. What will be the concentration of A after 7.5 minutes?

EK 4.A.1, 4.A.2
Student Edition
pp. 576–577

For the reaction above, how long will it take for 50% of the starting material to have been converted to B?

Section 13.4 Activation Energy and Temperature Dependence of Rate Constants, *pp. 588–593*

Essential Knowledge Covered:
4.B.1, 4.B.2, 4.B.3

REVIEW IT

EK 4.B.1, 4.B.2
Student Edition p. 588

What is the major issue with the idea that simple reaction kinetics of reactant particle collisions can be used to determine the rate of a reaction?

_____

_____

_____

_____

_____

_____

_____

EK 4.B.2
Student Edition
pp. 589–590

State the Arrhenius equation and explain the terms associated with the equation.

_____

_____

_____

USE IT

EK 4.B.2
Student Edition p. 590

When the Arrhenius equation is simplified using logarithms, it can be expressed as $ln\ k = \left(-\dfrac{E_a}{R}\right)\left(\dfrac{1}{T}\right) + ln\ A$. Explain how this can be used to create a graph that would illustrate a linear relation. Describe the linear relation that would result.

EK 4.B.1, 4.B.2, 4.B.3

SUMMARIZE IT

Draw potential energy profiles for endothermic and exothermic reactions. Fully label these profiles assuming that they describe the reaction of A + B forming C + D.

EK 4.B.2

Use the data below to determine the activation energy of the reaction from which the data was collected.

k	T (K)
0.014	800
0.038	840
0.109	880
0.375	920

Section 13.5 Reaction Mechanisms, *pp. 594–599*

Essential Knowledge Covered:
4.C.1, 4.C.2, 4.C.3

REVIEW IT

EK 4.C.1, 4.C.2, 4.C.3
Student Edition p. 595

Word Bank

determining
elementary
intermediate
mechanism

Using the word bank on the left, **fill in the blanks.** Words may be used more than once.

The balanced chemical reaction is often a summary of a series of _____ steps that occur as the reaction proceeds from reactants to products. This set of steps is called the reaction _____. Any species (molecule or particle that show up in a step of the reaction _____) that is not in the overall reaction is called a(n) _____. This always forms, but is later consumed as the reaction proceeds from reactants to products. The slowest step in any reaction mechanism is called the rate _____ step, as it is the step that controls the rate at which steps after it can occur.

EK 4.C.1, 4.C.2
Student Edition p. 595

What two conditions must the elementary steps in a chemical reaction satisfy?

_____

_____

_____

USE IT

EK 4.C.1, 4.C.2, 4.C.3
*Student Edition
pp. 596–597*

In the reaction $4HBr\,(g) + O_2(g) \rightarrow 2H_2O(g) + 2Br_2(g)$, it has been found experimentally that the change in initial concentration of the HBr has the same effect on the rate as a change in the initial concentration of O_2. It is also proposed that intermediates HOOBr and HOBr form in the mechanism. Suggest a possible mechanism for this reaction and include a reason as to why this process does not take place in a single step.

SUMMARIZE IT

EK 4.C.1, 4.C.2, 4.C.3

Fill in the sequence of steps used in the study of a reaction mechanism:

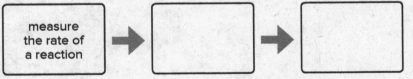

measure
the rate of
a reaction

EK 4.B.3, 4.D.1, 4.D.2
Student Edition p. 599

USE IT

A catalyst is a substance that can be added to a reaction that increases the rate by (i) decreasing the activation energy by providing an alternative pathway or mechanism, (ii) by changing the geometry of the collisions to make them more favorable, or (iii) a combination of the two. While it may play a role in the formation of an intermediate species, it is a substance that is never consumed in the reaction.

Create a sketch of a potential energy pathway for a non-catalyzed and a catalyzed exothermic reaction.

SUMMARIZE IT

EK 4.D.2

What are the three general types of catalysis?

_____

_____

EK 4.D.1, 4.D.2

Give an example of each type of catalyst and create a sketch to illustrate each process where possible.

These questions were posed in the *Chemistry* chapter opener (page 562). Answer them using the knowledge you've gained from this chapter.

1. What is a reaction rate? What is a rate constant? What is a rate law?

_____

_____

2. How can we determine the order of a reaction or the order with respect to a given reactant?

_____

_____

3. What is a half-life?

_____

_____

_____

4. How is the reaction rate related to collisions?

_____

_____

_____

5. What is a rate determining step?

_____

_____

_____

6. What is an intermediate?

_____

_____

7. What is a catalyst?

_____

_____

_____

8. What is a surface catalyst? What is an enzyme?

_____

_____

_____

As you work through your AP Focus Review Guide, keep this chapter's Big Ideas in mind:

AP A LOOK AHEAD

BIG IDEA
6
A system at equilibrium produces no observable changes over time, and this equilibrium can be manipulated in a predictable fashion.

Section 14.1 The Concept of Equilibrium and the Equilibrium Constant, *pp. 622–625*

Essential Knowledge Covered:
6.A.4

USE IT

Answer the question that follow based on the diagram below.

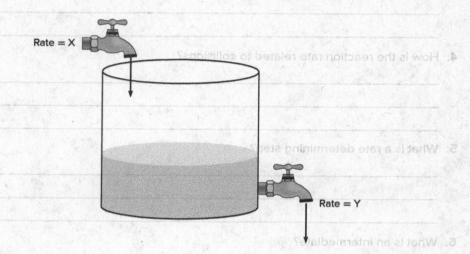

Rate = X

Rate = Y

EK 6.A.4
Student Edition p. 625

What will happen to the system if

1. X > Y: _____

_____

2. X < Y: _____

_____

3. X = Y: _____

_____

Section 14.1 The Concept of Equilibrium and the Equilibrium Constant (continued)

Which numbered case on page 142 is similar to a system at equilibrium? Why?

_____

_____

_____

_____

SUMMARIZE IT

EK 6.A.4

Complete the table summarizing the possible values of '*K*' and the meaning of their magnitudes:

If the values for *K* are:	Then the reaction favors:
$K \gg 1$	
$K = 1$	
$K \ll 1$	

Section 14.2 Writing Equilibrium Constant Expressions, *pp. 625–636*

Essential Knowledge Covered:
6.A.1, 6.A.2, 6.A.3, 6.A.4

REVIEW IT

EK 6.A.1, 6.A.2,
6.A.3, 6.A.4
Student Edition
pp. 624–627

Complete the following comparison charts that define heterogeneous and homogeneous equilibria, and how equilibrium constants are written.

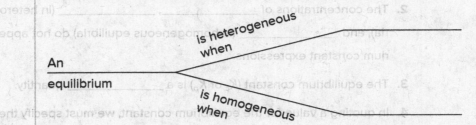

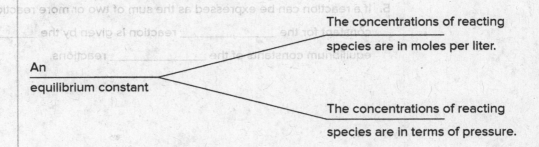

USE IT

EK 6.A.2
Student Edition p. 625

For the following reactions at equilibrium, determine the correct expression for the equilibrium constant.

$$HF(aq) + H_2O(l) \rightleftharpoons H_3O^+(aq) + F^-(aq)$$

$$N_2O_5(g) \rightleftharpoons NO_2(g) + O_2(g)$$

SUMMARIZE IT

EK 6.A.2, 6.A.3

Using the word bank on the left, **fill in the blanks** in the summary of guidelines for writing K_{eq} expressions below.

Word Bank

balanced equation
concentrations
dimensionless
individual
overall
product
pure liquids
pure solids
reacting
solvents
temperature

1. The _____ of the _____ species in the condensed phase are expressed in mol/L; in the gaseous phase, the concentrations can be expressed in mol/L or in atm. K_c is related to K_p by a simple equation ($K_p = K_c(RT)^{\Delta n}$.)

2. The concentrations of _____, _____ (in heterogeneous equilibria), and _____ (in homogeneous equilibria) do not appear in the equilibrium constant expressions.

3. The equilibrium constant (K_c or K_p) is a _____ quantity.

4. In quoting a value for the equilibrium constant, we must specify the _____ _____ and _____.

5. If a reaction can be expressed as the sum of two or more reactions, the equilibrium constant for the _____ reaction is given by the _____ of the equilibrium constants of the _____ reactions.

Section 14.3 The Relationship Between Chemical Kinetics and Chemical Equilibrium, *pp. 637–638*

Essential Knowledge Covered:
6.A.3

EK 6.A.3
Student Edition p. 637

REVIEW IT

Why is K_c always a constant at a given temperature regardless of the equilibrium concentrations of the reacting species?

_____

_____

EK 6.A.3
Student Edition p. 637

USE IT

If the equilibrium constant for a particular reaction is 4.8×10^{-2} at 80°C and the forward rate constant is 3.2×10^2 s^{-1}, what is the value of the reverse rate constant?

EK 6.A.3

SUMMARIZE IT

Why is K_c only constant at a given temperature?

_____

_____

_____

_____

Section 14.4 What Does the Equilibrium Constant Tell Us?, *pp. 638–644*

Essential Knowledge Covered:
6.A.1, 6.A.2, 6.A.3, 6.A.4

REVIEW IT

Word Bank

does not
equilibrium
large
left
products
reactants
right
small

Using the word bank on the left, **fill in the blanks** in the following discussion of the relationship between Q_c and K_c. Words may be used more than once or not at all.

- $Q_c < K_c$: The ratio of initial concentrations of products to reactants is too _____.
  The reaction must shift to the _____ to reach equilibrium, which corresponds to _____ being converted to products.

- $Q_c = K_c$: The initial concentrations are at the equilibrium concentrations. The reaction is at _____ and _____ move in either direction.

- $Q_c > K_c$: The ratio of initial concentrations of products to reactants is too _____.
  The reaction must shift to the _____ to reach equilibrium, which corresponds to _____ being converted to reactants.

EK 6.A.1
Student Edition p. 641

Circle the diagram below that represents a Q_c/K_c relationship that would lead to an increase in product concentration:

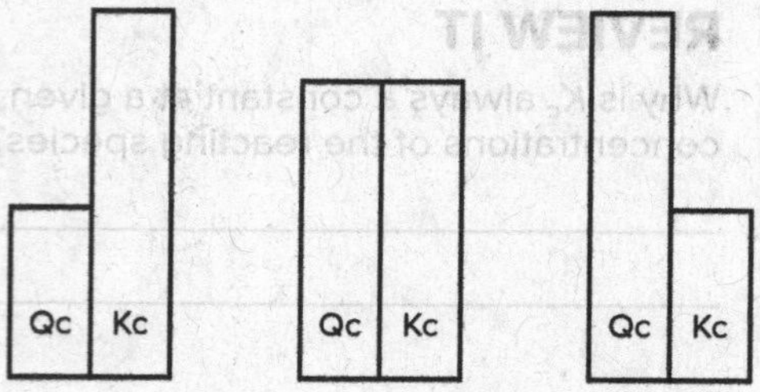

USE IT

EK 6.A.2 6.A.4
Student Edition p. 641

For the reaction $N_2O_4(g) \rightleftharpoons 2NO_2(g)$, $K_c = 0.21$ at 100°C. At a point during the reaction, $[N_2O_4] = 0.12$ M and $[NO_2] = 0.55$ M. Is the reaction at equilibrium at this point? If not, in which direction is it progressing?

EK 6.A.3 6.A.4
Student Edition p. 641

0.200 mol HI gas is placed in a 2.00 L flask and allowed to proceed according to the following reaction at 453 °C:

$2HI(g) \rightleftharpoons H_2(g) + I_2(g)$

Calculate K_c if [HI] = 0.078 M at equilibrium.

Section 14.4 What Does the Equilibrium Constant Tell Us? (continued)

SUMMARIZE IT

EK 6.A.2, 6.A.3

What would be the benefit of monitoring Q_c to a researcher who is investigating the kinetics of a particular system?

_____

_____

_____

Section 14.5 Factors That Affect Chemical Equilibrium, *pp. 644–654*

Essential Knowledge Covered:
6.A.2, 6.A.3, 6.B.1, 6.B.2

REVIEW IT

EK 6.B.1
Student Edition p. 644

Explain how a system at equilibrium's reaction to a stress can be predicted by LeChâtelier's principle.

_____

_____

_____

_____

USE IT

EK 6.B.2
Student Edition p. 645

Determine the effects when each of the suggested stresses are applied to the reaction:

$$2H_2S(g) + O_2(g) \rightleftharpoons 2S(s) + 2H_2O(g)$$

1. What happens to $[H_2O]$ if O_2 is added? _____

_____

2. What happens to $[O_2]$ if H_2S is removed? _____

_____

3. What happens to $[O_2]$ if the volume of the reactor is halved? _____

4. What happens to $[O_2]$ if a catalyst is added? _____

_____

SUMMARIZE IT

EK 6.B.1, 6.B.2

The Fischer Esterification is a reaction in Organic Chemistry by which an ester is synthesized from an alcohol and a carboxylic acid according to the following general scheme:

RCO_2H (a carboxylic acid) + $R'OH$ (an alcohol) $\rightleftharpoons$ RCO_2R' (an ester) + H_2O (l). Suggest a method by which a synthetic organic chemist could increase the yield of ester.

_____

_____

_____

AP Reviewing the Essential Questions

These questions were posed in the *Chemistry* chapter opener (page 621). Answer them using the knowledge you've gained from this chapter.

1. What is a reversible reaction?

_____

_____

2. What is the relationship between summed reactions and their individual equilibrium constants?

_____

_____

_____

3. What is an equilibrium constant?

_____

_____

4. What is Q? What does Q indicate about a reaction returning to equilibrium? What is an ICE table?

_____

_____

_____

5. Given a reaction and its equilibrium constant can we calculate reactant or product concentrations?

_____

_____

_____

6. What is Le Chatelier's principle?

_____

_____

As you work through your AP Focus Review Guide, keep this chapter's Big Ideas in mind:

AP **A LOOK AHEAD**

BIG IDEA
6
Acid-base dynamics can be understood in terms of intermolecular forces, bond behavior, and dynamic equilibriums.

Section 15.1   Brønsted Acids and Bases, *pp. 667–668*

Essential Knowledge Covered:
6.C.1

EK 6.C.1
Student Edition
pp. 667–668

REVIEW IT

Identify the two conjugate acid-base pairs in the following reaction. Be sure to indicate which species in each pair is the acid and which is the base.

$$HSO_3^- \,(aq) + CH_3COO^- \,(aq) \rightleftharpoons SO_3^{2-} \,(aq) + CH_3COOH \,(aq)$$

_____

_____

_____

Section 15.2   The Acid-Base Properties of Water, *pp. 668–670*

Essential Knowledge Covered:
6.C.1, 6.C.2

EK 6.C.1, 6.C.2
Student Edition
pp. 669–670

USE IT

Determine the concentration of the hydroxide ions in solution at 25°C for each solution below (assume that each acid given fully dissociates):

0.125 M solution of hydrochloric acid

2.75×10^{-4} M solution of nitric acid

EK 6.C.1, 6.C.2
Student Edition
pp. 669–670

Determine the concentration of H^+ ions in each solution at 25°C.

A solution with a hydroxide concentration of 1.85×10^{-6} M

A solution with a hydroxide concentration of 4.50 M

Section 15.3   pH- A Measure of Acidity, *pp. 670–673*

Essential Knowledge Covered:
6.C.1, 6.C.2

SUMMARIZE IT

EK 6.C.1, 6.C.2

Complete the flowchart below that can be used to determine the pH of a solution where you are given information regarding the number of grams of solid sodium hydroxide that was placed in water to create a known volume of a base solution. Is there more than one possibility for this flowchart? Explain.

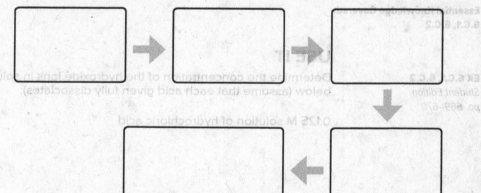

EK 6.C.1, 6.C.2

Use your flowchart to determine the pH of a solution formed by placing 1.00 g of sodium hydroxide in enough water to make a 250 mL base solution.

Na: 22.99 g/mol
O: 16.00 g/mol
H: 1.01 g/mol

Section 15.4 Strengths of Acids and Bases, *pp. 673–677*

Essential Knowledge Covered:
6.C.1, 6.C.2

USE IT

EK 6.C.1
Student Edition p. 675

In the ionization of ammonia $NH_3\ (aq) + H_2O(l) \rightleftharpoons NH_4^+\ (aq) + OH^-\ (aq)$, why would the NH_3 be considered a Brønsted base?

EK 6.C.1
Student Edition p. 677

Use the chart given here to predict the direction of the given reaction in aqueous solution.

Acid	Conjugate Base
$HClO_4$ (perchloric acid)	ClO_4^- (perchlorate ion)
HI (hydroiodic acid)	I^- (iodide ion)
HBr (hydrobromic acid)	Br^- (bromide ion)
HCl (hydrochloric acid)	Cl^- (chloride ion)
H_2SO_4 (sulfuric acid)	HSO_4^- (hydrogen sulfate ion)
HNO_3 (nitric acid)	NO_3^- (nitrate ion)
H_3O^+ (hydronium ion)	H_2O (water)
HSO_4^- (hydrogen sulfate ion)	SO_4^{2-} (sulfate ion)
HF (hydrofluoric acid)	F^- (fluoride ion)
HNO_2 (nitrous acid)	NO_2^- (nitrite ion)
HCOOH (formic acid)	$HCOO^-$ (formate ion)
CH_3COOH (acetic acid)	CH_3COO^- (acetate ion)
NH_4^+ (ammonium ion)	NH_3 (ammonia)
HCN (hydrocyanic acid)	CN^- (cyanide ion)
H_2O (water)	OH^- (hyroxide ion)
NH_3 (ammonia)	NH_2^- (amide ion)

Strong acids / Weak acids

Acid strength increases

Base strength increases

Section 15.4   Strengths of Acids and Bases (continued)

$$HF\ (aq) + COOH^-\ (aq) \rightleftharpoons HCOOH\ (aq) + F^-\ (aq)$$

_____

_____

_____

_____

SUMMARIZE IT

EK 6.C.1, 6.C.2

Create a sketch of a fully ionized strong acid and a partially ionized weak acid.

Section 15.5   Weak Acids and Acid Ionization Constants, *pp. 677–685*

Essential Knowledge Covered:
6.C.1, 6.C.2

USE IT

EK 6.C.1, 6.C.2
Student Edition p. 683

Determine the K_a of a weak acid where the pH of a 0.12 M HA solution is 3.16.

EK 6.C.1
Student Edition p. 684

What is the percent ionization of the acid in the previous question?

EK 6.C.1, 6.C.2

SUMMARIZE IT

Create a flowchart to identify the steps in solving for the pH of a weak acid ionization problem.

EK 6.C.1, 6.C.2

Use your flow chart to determine the pH of a 0.25 M hydrofluoric acid solution ($K_a = 7.1 \times 10^{-4}$).

Section 15.6   Weak Bases and Base Ionization Constants, *pp. 685–687*

Essential Knowledge Covered:
6.C.1, 6.C.2

EK 6.C.1, 6.C.2
Student Edition p. 685

USE IT

Calculate the pH of a 0.25 M pyridine solution where the K_b value is 1.7×10^{-9}. The reaction for the ionization is $C_5H_5N\ (aq) + H_2O(l) \rightleftharpoons C_5H_5NH^+\ (aq) + OH^-\ (aq)$

EK 6.C.1
Student Edition p. 685

Why is water not included in the K_b expression in the reaction in the previous question?

_____

_____

Section 15.7   The Relationship Between the Ionization Constants of Acids and Their Conjugate Bases, *pp. 687–688*

Essential Knowledge Covered:
6.C.1, 6.A.4

6.A.4
Student Edition p. 687

REVIEW IT

What is true for all conjugate acid-base ionization constants?

_____

Section 15.7   The Relationship Between the Ionization Constants of Acids and Their Conjugate Bases (continued)

EK 6.C.1, 6.A.4

SUMMARIZE IT

Given the following K_b values for four weak bases, order their conjugate acids in increasing strength. Include a calculation of each K_a value.

Weak base	K_b
Ammonia	1.8×10^{-5}
Aniline	3.8×10^{-10}
Methylamine	4.4×10^{-4}
Urea	1.5×10^{-14}

Section 15.8   Diprotic and Polyprotic Acids, *pp. 688–692*

Essential Knowledge Covered:
6.C.1, 6.C.2, 6.A.4

EK 6.C.1
Student Edition p. 689

USE IT

Which step in the ionization of a polyprotic acid is responsible for the pH of the solution? Explain your answer.

_____

_____

_____

EK 6.C.1, 6.C.2, 6.A.4
Student Edition
pp. 689–691

What would be the pH of a 0.10 M carbonic acid solution, based on the given information:

$$K_{a1} = 4.2 \times 10^{-7}, \quad K_{a2} = 4.8 \times 10^{-11}$$

SUMMARIZE IT

EK 6.C.1, 6.C.2

Write the ionizations for citric acid $H_3C_6H_5O_7$(aq).

First ionization: _____

Second ionization: _____

Third ionization: _____

For each ionization in the previous question, write the acid ionization constants.

Section 15.9 Molecular Structure and the Strength of Acids, *pp. 692–696*

Essential Knowledge Covered:
6.C.1

USE IT

EK 6.C.1
Student Edition p. 695

Put the following acids in decreasing strengths: H_3PO_2 (aq), H_3PO_4 (aq), H_3PO (aq) and H_3PO_3 (aq). Explain your answer.

____

____

____

SUMMARIZE IT

EK 6.C.1

Create a flow chart to organize *binary acids*, *oxyacids* and *carboxylic acids* in terms of how to determine if the acid is strong or weak.

EK 6.C.1, 6.C.2
Student Edition
pp. 697–698

USE IT

What is the pH of a 0.19 M potassium acetate solution?

EK 6.C.1
Student Edition p. 698

Determine the percent hydrolysis of the previous solution.

EK 6.C.1, 6.C.2, 6.A.4

SUMMARIZE IT

Create a graphic organizer of your choice to help organize the process in which a salt is present in solution where both the cation and the anion hydrolyze.

Use the headings: If $K_b > K_a$, If $K_b < K_a$, and If $K_b \approx K_a$

Section 15.11 Acid-Base Properties of Oxides and Hydroxides, *pp. 702–704*

Essential Knowledge Covered:
6.C.1, 6.C.2

REVIEW IT

EK 6.C.1
Student Edition p. 703

Identify the salt that will be created in each reaction:

$CO_2(g) + 2\ KOH(aq)$:

_____

$CaO + 2\ HCl$:

_____

SUMMARIZE IT

EK 6.C.1

Complete the following reactions:

$SrO(s) \xrightarrow{H_2O}$

$SO_2(g) + H_2O(l) \rightarrow$

$N_2O_5(g) + H_2O(l) \rightarrow$

EK 6.C.1, 6.C.2

Explain how carbonating water to create a soda results in the formation of an acidic solution.

_____

_____

_____

_____

Section 15.12 Lewis Acids and Bases, *pp. 704–708*

Prerequisite Knowledge

This section covers concepts you should know from other science courses. Review it to master the AP Essential Knowledge and prepare for the AP Exam.

SUMMARIZE IT

Compare and contrast a Lewis acid and base with a Brønsted acid and base.

_____

_____

_____

_____

These questions were posed in the *Chemistry* chapter opener (page 666). Answer them using the knowledge you've gained from this chapter.

1. What is a Brønsted acid? What is a Brønsted base?

   _____

   _____

   _____

2. What are pH and pOH?

   _____

   _____

   _____

3. What is a strong acid? What is a strong base? What is a weak acid? What is a weak base?

   _____

   _____

   _____

4. What is the relationship between Ka and Kb for a conjugate acid-base pair?

   _____

   _____

   _____

5. How does molecular structure influence acid strength?

   _____

   _____

   _____

6. When a salt of a weak acid hydrolyzes, is the anion produced a proton acceptor or donor?

   _____

   _____

   _____

7. What are the expected pH ranges when metal oxides or non-metal oxides react with water? What are amphoteric compounds?

   _____

   _____

   _____

   _____

ACID-BASE EQUILIBRIA AND SOLUBILITY EQUILIBRIA

As you work through your AP Focus Review Guide, keep this chapter's Big Ideas in mind:

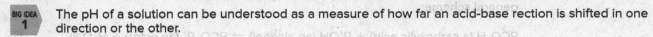

AP **A LOOK AHEAD**

BIG IDEA 1 The pH of a solution can be understood as a measure of how far an acid-base rection is shifted in one direction or the other.

BIG IDEA 3 Acid-base reactions can be viewed as proton-transfer events with the transfer occuring from acid to base.

BIG IDEA 6 The rules of equilibria readily apply to acid-base reactions.

Section 16.1 Homogeneous versus Heterogeneous Solution Equilibria, *p. 721*

Essential Knowledge Covered:
6.C.1, 6.C.3

EK 6.C.1, 6.C.3
Student Edition p. 721

REVIEW IT

There are two diagrams below. Identify which diagram represents homogeneous solution equilibrium and which diagram represents heterogeneous solution equilibrium.

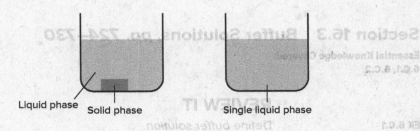

Liquid phase   Solid phase   Single liquid phase

SUMMARIZE IT

EK 6.B.1, 6.B.2, 6.B.4

The Fischer Esterification is a reaction in organic chemistry by which an ester is synthesized from an alcohol and a carboxylic acid according to the following general scheme:

RCO_2H (a carboxylic acid) $+ R'OH$ (an alcohol) $\rightleftharpoons RCO_2R'$ (an ester) $+ H_2O$ (l).

How would the yield of this reaction be affected by the addition of a large amount of ester (RCO_2R') to the reaction vessel?

_____

_____

_____

How is this example similar to the common ion effect?

_____

_____

Section 16.3   Buffer Solutions, *pp. 724–730*

Essential Knowledge Covered:
6.C.1, 6.C.2

REVIEW IT

EK 6.C.1
Student Edition p. 724

Define *buffer solution*.

_____

_____

Select which combinations below would create a buffer system

a. KCN/HCN
b. Na_2SO_4/$NaHSO_4$
c. NH_3/NH_4NO_3
d. NaI/HI
e. KCl/HCl

EK 6.C.1, 6.C.2
Student Edition
pp. 725–730

USE IT

What is the pH of a buffer system that contains 0.50 M acetic acid and 0.50 M sodium acetate? How does the pH of this solution change when 0.10 mol gaseous HCl is added to the solution (assume the volume does not change?)

**EK 1.E.2, 3.A.2,
3.B.2, 6.C.1**

SUMMARIZE IT

Calculate the pH during the titration of 40.00 mL of 0.1000 M propanoic acid (HPr, $K_a = 1.3 \times 10^{-5}$) after the addition of 30.00 mL of a solution that is 0.1000 M in NaOH.

Essential Knowledge Covered:
3.B.2, 6.C.1

REVIEW IT

EK 3.B.2
Student Edition p. 739

Given the following titration curves, use Table 16.1 on page 741 of your textbook to select an indicator for each titration. The equivalence point is marked with an 'X.'

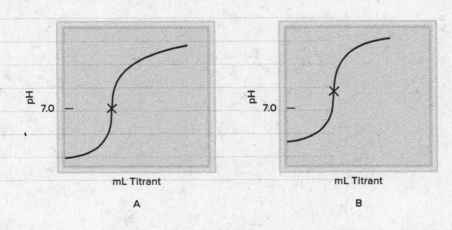

A

B

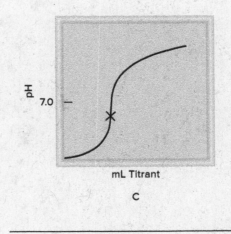

C

EK 6.C.1
Student Edition p. 742

SUMMARIZE IT

A student performing a titration gradually adds NaOH solution to a solution of HCl. Upon nearing the equivalence point, she observes a faint pink color (the selected indicator is phenolphthalein). The color, however, fades quickly. What should the student do next? What happened in the flask? How could this be prevented?

_____

_____

_____

_____

_____

_____

_____

_____

EK 6.C.3

SUMMARIZE IT

If 2.00 mL of 0.200 M NaOH are added to 1.00 L of 0.100 M $CaCl_2$, will precipitation occur? K_{sp} for the dissociation of $Ca(OH)_2$ is 8.0×10^{-6}.

Essential Knowledge Covered:
6.B.1, 6.C.3

REVIEW IT

EK 6.B.1, 6.C.3
Student Edition p. 749

Given the table of K_{sp} values on page 745, select an appropriate reagent that would enable fractional precipitation of one anion in each pair, then identify the compound that will precipitate first:

Anion pair	Precipitating cation	Compound to precipitate first
Br^- vs I^-		
OH^- vs CO_3^{2-}		
CO_3^{2-} vs SO_4^{2-}		
Cl^- vs S^{2-}		
Cl^- vs CrO_4^{2-}		

SUMMARIZE IT

EK 6.C.3

A solution contains 0.20 M $MgCl_2$ and 0.10 M $CuCl_2$. Calculate the concentration of hydroxide that would separate the metal ions. The relevant values for K_{sp} are:

$Mg(OH)_2 = 6.3 \times 10^{-10}$
$Cu(OH)_2 = 2.2 \times 10^{-20}$

Essential Knowledge Covered:
6.B.1, 6.B.2, 6.C.3

REVIEW IT

EK 6.B.1, 6.B.2, 6.C.3
Student Edition p. 751

In the diagram below, fill in the boxes with the appropriate species.

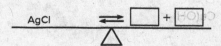

How would the solubility of AgCl be affected by the addition of AgNO₃?

_____

_____

How is the diagram of the scale appropriate to the study of solubility equilibria?

_____

_____

_____

SUMMARIZE IT

EK 6.C.3

The solubility of $Ca(OH)_2$ in water is 1.2×10^{-2} M.

Would the solubility of $Ca(OH)_2$ be greater or less than 1.2×10^{-2} M in a solution of 0.10 $Ca(NO_3)_2$?

_____

_____

_____

Would the solubility of $Ca(OH)_2$ increase or decrease at high pH relative to low pH?

_____

_____

_____

Essential Knowledge Covered:
6.B.1

EK 6.B.1
Student Edition p. 756

USE IT

Determine whether the following compounds will be more soluble in acidic solution than in water, and support your answer with the appropriate reactions.

$Ca(OH)_2$

_____

_____

$Mg_3(PO_4)_2$

_____

_____

$PbBr_2$

_____

_____

_____

Section 16.10   Complex Ion Equilibria and Solubility, *pp. 756–761*

Essential Knowledge Covered:
6.B.1, 6.B.2, 6.C.3

REVIEW IT

Using the word bank on the left, **fill in the blanks.**

Word Bank

complex ion
equilibrium
formation constant
increase
ions
Lewis acid
Lewis base
molecules
stability constant
transition

A _____ is an ion formed when a metal cation reacts with a _____ and defined generally as ions that contain a central metal cation that is bonded to one or more _____ or _____. These bonded species are referred to in general as ligands. While the chemistry of complex ions is dealt with later, their formation has important implications in terms of solubility. _____ metals are especially likely to form complex ions, as they have unfilled *d*-type orbitals that can accept electron density, thus behaving as _____. The ability of a metal to form a complex ion will generally _____ the solubility of the compound that contains that cation, because when the complex ion forms, that cationic species is removed from solution, at which point the _____ will shift accordingly. The tendency for a metal to form complex ions is quantitated by the _____, K_f, which is the equilibrium constant for the complex ion formation reaction. The formation constant is also known as the _____, because compounds with larger values for K_f tend to be more stable.

USE IT

EK 6.C.3
Student Edition p. 759

Paying particular attention to table 16.4 on page 759 of your textbook, how would each of the following compounds affect the solubility of CdS?

$LiNO_3$: _____

_____

_____

Na_2SO_4: _____

_____

_____

KCN: _____

_____

_____

$NaClO_4$: _____

_____

_____

KI: _____

_____

_____

EK 6.B.1, 6.B.2

SUMMARIZE IT

Define the terms *Lewis acid* and *Lewis base*:

Lewis acid: _____ or _____

Lewis base: _____

Determine the identity of the Lewis acid and the Lewis base in each of the following reactions:

Reaction	Lewis acid	Lewis base
$C_2H_5OC_2H_5 + AlCl_3 \rightarrow (C_2H_5)_2OAlCl_3$		
$Hg^{2+} + 4CN^- \rightarrow Hg(CN)_4^{2-}$		
$Ag^+ + 2NH_3 \rightarrow Ag(NH_3)_2^+$		
$H_2O + CO_2 \rightarrow CH_2O_3$		
$H_2O + CO_2 \rightarrow CH_2O_3$		

LiNO₃

Na₂SO₄

KCN

NaClO₄

Section 16.11 Application of the Solubility Product Principle to Qualitative Analysis, *pp. 761–763*

Essential Knowledge Covered:
6.C.3

Word Bank

aluminum
barium
bismuth
cadmium
calcium
chromium
cobalt
copper
iron
lead
manganese
mercury(I)
mercury(II)
nickel
silver
strontium
tin
zinc

REVIEW IT

Place the ions in the word bank into the appropriate box that identifies the point in the qualitative analysis process where that ion would be removed from solution.

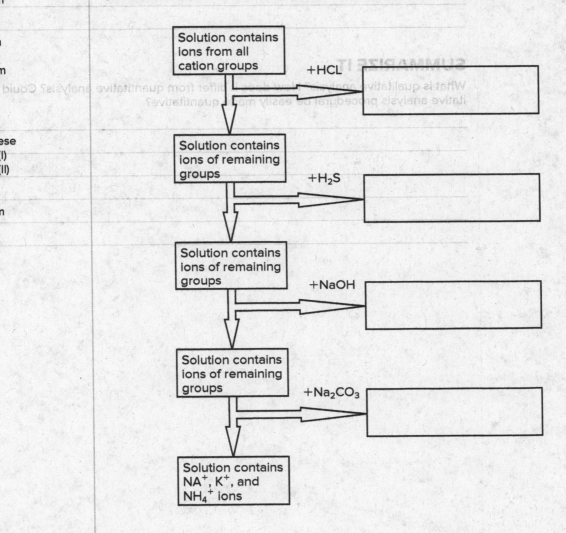

EK 6.C.3
Student Edition p. 764

USE IT

How could a mixture of lead and copper be separated qualitatively?

_____

_____

EK 6.C.3

SUMMARIZE IT

What is qualitative analysis? How does it differ from quantitative analysis? Could a qualitative analysis procedural be easily made quantitative?

_____

_____

_____

_____

_____

These questions were posed in the *Chemistry* chapter opener (page 720). Answer them using the knowledge you've gained from this chapter.

1. What is the common ion effect?

_____

_____

2. What is a buffer?

_____

_____

3. What is acid-base titration?

_____

_____

4. What is K_{sp}? How does the magnitude of K_{sp} help identify the order in which ions can be separated from a solution through precipitation techniques?

_____

_____

_____

5. How does pH affect solubility?

_____

_____

_____

Chapter 17

ENTROPY, FREE ENERGY, AND EQUILIBRIUM

As you work through your AP Focus Review Guide, keep this chapter's Big Ideas in mind:

AP   A LOOK AHEAD

BIG IDEA 5   Energy and entropy determine the direction of a chemical reaction.

BIG IDEA 6   Equilibrium states exist based on intermolecular and bonding forces.

Section 17.1   The Three Laws of Thermodynamics, *p. 777*

Essential Knowledge Covered:
5.B.2

REVIEW IT

EK 5.B.2
Student Edition p. 777

There are _____ laws of thermodynamics.

What is the first law of thermodynamics?

_____

_____

What does the second law of thermodynamics help to explain?

_____

_____

_____

Section 17.2   Spontaneous Processes, *pp. 777–778*

Essential Knowledge Covered:
5.E.2, 5.E.5

REVIEW IT

EK 5.E.2, 5.E.5
Student Edition p. 777

What is meant by the term *spontaneous process* (or thermodynamically favored process) in chemistry?

_____

_____

EK 5.E.5
Student Edition p. 777

Give two examples of physical processes that occur spontaneously and two chemical processes that occur spontaneously.

_____

_____

_____

_____

_____

Section 17.2   Spontaneous Processes (continued)

USE IT

EK 5.E.5
Student Edition p. 778

Besides enthalpy, what is the name of the thermodynamic quantity that must be considered when determining the spontaneous direction of a chemical reaction?

_____

EK 5.E.2, 5.E.5
Student Edition p. 778

Is it possible for a reaction to be spontaneous in the endothermic direction? Explain your answer.

_____

_____

_____

Section 17.3   Entropy, *pp. 778–783*

Essential Knowledge Covered:
5.E.1

REVIEW IT

Using the word bank on the left, **fill in the blanks.**

Word Bank

entropy
rotational
translational
vibrational

The *measure* of how dispersed the energy of a system is among the different possible

ways that a system can contain energy is referred to as _____. Most chemical

processes are accompanied by a change in this quantity. Examples of the energy

states of a system include _____, _____ and _____

motions of the particles in the system.

EK 5.E.1
Student Edition p. 779

Define *microstate* and explain how this term applies to entropy.

_____

_____

USE IT

EK 5.E.1
Student Edition p. 780

What does the number of microstates tell us about the probability of a particular distribution existing in a system?

_____

_____

EK 5.E.1
Student Edition p. 780

If $W_i > W_f$, does the entropy increase or decrease? Explain your answer.

_____

_____

_____

Section 17.3   Entropy (continued)

SUMMARIZE IT

EK 5.E.1

What factors must be considered when looking qualitatively at the entropy change for the ionization of a salt in water?

_____

_____

_____

_____

EK 5.E.1

Use the table for standard entropy values below to determine if entropy increases or decreases when solid iodine sublimates. Explain your answer.

Standard Entropy Values ($S°$) for Some Substances at 25°C	
Substance	**$S°$ (J/K · mol)**
$H_2O(l)$	69.9
$H_2O(g)$	188.7
$Br_2(l)$	152.3
$Br_2(g)$	245.3
$I_2(s)$	116.7
$I_2(g)$	260.6
C (diamond)	2.4
C (graphite)	5.69
CH_4 (methane)	186.2
C_2H_6 (ethane)	229.5

Section 17.4　The Second Law of Thermodynamics, *pp. 783–788*

Essential Knowledge Covered:
5.E.1, 5.E.2

EK 5.E.1, 5.E.2
Student Edition p. 783

REVIEW IT

Write the three equations for the second law of thermodynamics for:
A spontaneous/thermodynamically favored process

_____

A non-spontaneous/thermodynamically non-favored process

_____

An equilibrium process

_____

USE IT

EK 5.E.1, 5.E.2
Student Edition p. 784

Use the standard entropy values below to calculate the standard entropy changes for the reaction

$$CH_4 (g) + 2O_2 (g) \rightarrow CO_2 (g) + 2H_2O (g)$$

Molecule	Standard entropy (J/K · mol)
CH_4	186.2
O_2	205.0
CO_2	213.6
H_2O	188.7

EK 5.E.1, 5.E.2
Student Edition p. 785

Predict whether the entropy change of the system in each reaction is positive or negative. Explain your answer.

$$2SO_3 (g) \rightarrow 2SO_2 (g) + O_2 (g)$$

_____

_____

$$NH_4^+ (aq) + NO_3^- (aq) \rightarrow NH_4NO_3 (s)$$

_____

_____

SUMMARIZE IT

EK 5.E.1, 5.E.2

For the reaction of $X_2 + Y_2$ to form X_2Y_3:

Write a balanced chemical equation for this reaction.

_____

Draw a diagram of the particles both before and after the reaction.

_____

What is the sign of ΔS for the reaction? Explain your answer.

_____

EK 5.E.1, 5.E.2

State the third law of thermodynamics.

_____

What is the concept of the *true* value of absolute entropy?

_____

Section 17.5   Gibbs Free Energy, *pp. 789–796*

Essential Knowledge Covered:
5.E.2, 5.E.3, 5.E.4

EK 5.E.3
Student Edition p. 789

REVIEW IT

What is the equation for Gibbs free energy?

_____

_____

_____

EK 5.E.2, 5.E.3, 5.E.4
Student Edition
pp. 795–796

USE IT

The molar heats of fusion and vaporization of a solid are 12.4 kJ/mol and 33.8 kJ/mol respectively. Calculate the entropy changes for the transitions of solid → liquid and for liquid → vapor for this material. At 1.0 atm, this compound melts at 6.7°C and boils at 91.2°C.

EK 5.E.2, 5.E.3, 5.E.4
Student Edition p. 796

Explain how the answer to the question above illustrates that $\Delta S_{vap} > \Delta S_{fus}$.

_____

_____

_____

Section 17.5   Gibbs Free Energy (continued)

SUMMARIZE IT

EK 5.E.2, 5.E.3, 5.E.4

What are the conditions for the possible types of reactions in terms of the change in free energy?

_____

_____

_____

_____

EK 5.E.3

What is the general equation for the standard free energy change for a reaction?

EK 5.E.2, 5.E.3, 5.E.4

Create a graphic organizer (such as a table or flow chart) to aid in determining the factors that affect the sign of ΔG in the relationship $\Delta G = \Delta H - T\Delta S$.

Section 17.6   Free Energy and Chemical Equilibrium, *pp. 796–800*

Essential Knowledge Covered:
6.D.1

USE IT

EK 6.D.1
Student Edition p. 796

What is the correction factor in the change in free energy equation when reactants and products are not at their standard states?

_____

_____

_____

EK 6.D.1
Student Edition p. 797

If at equilibrium, $\Delta G = 0$, what does Q become in the equation $\Delta G = \Delta G° + RT\, lnQ$?

_____

_____

What does the equation from above simplify to when at equilibrium? Show the steps in this simplification.

Why is this such a useful equation for scientists?

_____

_____

SUMMARIZE IT

EK 6.D.1

Fill in the following chart relating K and $\Delta G°$.

K	*ln* K	$\Delta G°$	Meaning
>1		negative	
= 1		zero	
< 1		positive	

Section 17.6 Free Energy and Chemical Equilibrium (continued)

EK 6.D.1

The equilibrium constant for the reaction of $A\ (g) + 2B\ (g) \rightarrow C\ (g)$ is 0.214 at 302 K. What is the standard free energy change for this reaction at this temperature, in kJ/mol?

EK 6.D.1

In a certain experiment, the initial pressures for this reaction are $P_A = 0.202$ atm, $P_B = 0.412$ atm and $P_C = 0.781$ atm. Calculate ΔG for the reaction at these pressures and predict the direction of the net reaction toward equilibrium at 302 K.

Section 17.7 Thermodynamics in Living Systems, *pp. 800–802*

Essential Knowledge Covered:
5.E.4

EK 5.E.4
Student Edition p. 800

REVIEW IT

Explain what occurs in a coupled reaction.

_____

_____

Section 17.7   Thermodynamics in Living Systems (continued)

EK 5.E.4
Student Edition p. 801

What is the role of an enzyme in a biological process?

_____

_____

Where does most of the free energy go as an enzyme acts in a biological system?

_____

_____

_____

EK 5.E.4

SUMMARIZE IT

Create a schematic representation of the ATP synthesis process and coupled reactions in living systems as amino acids are converted to proteins.

EK 5.E.4

Explain your schematic representation.

_____

_____

_____

_____

_____

Reviewing the Essential Questions

These questions were posed in the *Chemistry* chapter opener (page 776). Answer them using the knowledge you've gained from this chapter.

1. Can energy be destroyed or transferred between systems?

_____

_____

2. What is entropy?

_____

_____

3. Is an increase in the entropy of the universe thermodynamically favored?

_____

_____

_____

4. How is it possible for biological reactions with $\Delta G° > 0$ to occur?

_____

_____

_____

ELECTROCHEMISTRY

As you work through your AP Focus Review Guide, keep this chapter's Big Ideas in mind:

AP A LOOK AHEAD

BIG IDEA 3 Electrons can be seen as valid reactants or products in electrochemical reactions.

BIG IDEA 5 The energetic favorability of certain electrolytic processes can be exploited by coupling an energetically unfavorable process to them.

Section 18.1 Redox Reactions, *pp. 813–816*

Essential Knowledge Covered:
3.B.3, 3.C.3

USE IT

EK 3.B.3
Student Edition p. 815

Balance the following redox reaction which takes place in acidic media.

$$Fe^{2+} + MnO_4^- \rightarrow Fe^{3+} + Mn^{2+}$$

A: _____

B: _____

C: _____

D: _____

_____

_____

_____

SUMMARIZE IT

EK 3.C.3

How does electrochemistry convert chemical energy into electrical energy?

_____

_____

_____

_____

_____

Essential Knowledge Covered:
3.C.3

USE IT

EK 3.C.3
Student Edition p. 817

Identify parts A, B, C, and D of the following Galvanic cell and defend your selection of anode and cathode.

Current flows L to R

A: _____

B: _____

C: _____

D: _____

_____
_____
_____

SUMMARIZE IT

EK 3.C.3

You are designing a galvanic cell, and have the choice of several electrolytes for the salt bridge. If you are building a traditional Daniell-type galvanic cell (where the redox pair is Cu/Zn), which salt from the list below would make a good choice for the salt bridge? Which salt would make a poor choice? Why?

KCl, NH_4NO_3, $ZnSO_4$, $CuSO_4$, $ZnCl_2$, $CuCl_2$

_____
_____
_____
_____

Section 18.3 Standard Reduction Potentials, *pp. 818–824*

Essential Knowledge Covered:
3.C.3

REVIEW IT

Using the word bank on the left, **fill in the blanks** in the Standard Reduction Potential rules.

Word Bank

forward
intensive properties
negative
non-spontaneously
positive
reverse
reversed
reversible
spontaneously

1. The E° values apply to the half-cell reaction as read left-to-right, which is assumed to be the _____ direction.

2. The more _____ the value of E°, the greater the tendency for the substance to be reduced.

3. The half-cell reactions are _____.

4. Under standard-state conditions, any species on the left of a half-cell reaction will react _____ with a species that appears on the right of any half-cell reaction located below it on the table.

5. Changing the stoichiometric coefficients of a half-cell reaction does not affect the value of E° because electrode potentials are _____.

6. When a reaction is reversed, the magnitude remains the same but the charge is _____.

USE IT

EK 3.C.3
Student Edition
pp. 820–822

A voltaic cell is constructed to harness the electrochemical power of the following equation:

$Cl_2 (g) + Zn (s) \rightarrow Zn^{2+}(aq) + 2 Cl^- (aq)$ $E°_{cell} = 2.12$ V

Calculate $E°_{chlorine}$ given that $E°_{zinc} = -0.76$ V

Section 18.4   Thermodynamics of Redox Reactions, *pp. 824–827*

Essential Knowledge Covered:
3.C.3

EK 3.C.3
Student Edition p. 824

REVIEW IT

Fill in the table below.

ΔG°	K	E°_{cell}	Reactants or products favored?
negative	> 1		favors products
0		0	
positive			

EK 3.C.3

SUMMARIZE IT

Complete the following chart with the equations that relate all three fundamental quantities.

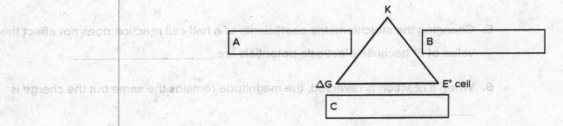

Extending Knowledge

This section takes the AP Essential Knowledge you have learned further, and may provide illustrative examples useful for the AP Exam.

USE IT

Student Edition p. 828

Predict whether the following reaction will occur spontaneously at 298 K with the given concentrations:

$Cd (s) + Fe^{2+} (aq) \rightarrow Cd^{2+} (aq) + Fe (s)$

$[Fe^{2+}] = 0.60$ M

$[Cd^{2+}] = 0.010$ M

Essential Knowledge Covered:
3.C.3

EK 3.C.3
Student Edition
pp. 832–833

REVIEW IT

Provide examples of the anodic, cathodic, and overall reactions for each of the 5 main types of battery:

Dry Cell:

Anode:
Cathode:
Overall:

Lithium Ion:

Anode:
Cathode:
Overall:

Fuel Cell:

Anode:
Cathode:
Overall:

USE IT

EK 3.C.3
Student Edition p. 835

How does a fuel cell differ from a traditional galvanic cell in the area of reactant handling?

_____

_____

_____

EK 3.C.3
Student Edition p. 833

When a lead storage battery is recharged, what are the redox reactions that are active? How do they differ from the reactions that are active when the battery is discharging?

_____

_____

_____

_____

_____

SUMMARIZE IT

EK 3.C.3

With their innocuous environmental profiles, why don't fuel cells currently see more widespread usage?

_____

_____

_____

_____

Section 18.7   Corrosion, *pp. 838–841*

Essential Knowledge Covered:
3.B.3

REVIEW IT

EK 3.B.3
Student Edition
pp. 839–840

Corrosion is, by definition, the degradation of a metal by an electrochemical process. Using iron as an example, use the principles of the galvanic cell to explain the following observations.

Iron requires moisture to rust.

_____

_____

Iron requires dissolved oxygen in water to rust.

_____

_____

The rate of rust formation increases as pH decreases.

_____

_____

USE IT

EK 3.B.3
Student Edition p. 839

If aluminum has a more negative SRP than iron, why does it tend to oxidize less readily?

_____

_____

EK 3.B.3
Student Edition p. 841

If iron rusts faster when it acts more readily as an anode, how can the rate of rust formation be slowed? Give two examples of how iron is protected from rusting in the developed world.

_____

_____

_____

_____

_____

_____

SUMMARIZE IT

EK 3.B.3

Why do cars that are driven in snowy climates tend to experience more problems with rust than cars that are driven in warmer climates?

_____

_____

_____

_____

EK 3.B.3

The process of 'galvanization' uses a layer of zinc to protect iron. Suggest another metal that could be used to protect iron.

_____

_____

Section 18.8   Electrolysis, *pp. 841–848*

Essential Knowledge Covered:
3.C.3, 5.E.4

REVIEW IT

EK 5.E.4
Student Edition p. 841

Define the terms below.

electrolysis: _____

_____

electrolytic cell: _____

_____

overvoltage: _____

_____

USE IT

EK 3.C.3
Student Edition p. 847

A technician plates a faucet with 0.84 g of Cr metal by electrolysis of aqueous $Cr_2(SO_4)_3$. If 12.5 minutes are allowed for the plating, what current is needed?

SUMMARIZE IT

EK 5.E.4

How is charging a dead car battery an example of an electrolytic process? Give the reaction that is occurring when a car battery is charged.

_____

_____

_____

_____

_____

EK 3.C.3

In an electrolytic process, what determines the amount of product that forms or the amount of reactant that is consumed?

_____

_____

_____

These questions were posed in the *Chemistry* chapter opener (page 812). Answer them using the knowledge you've gained from this chapter.

1. What is a redox reaction? What is oxidation? What is reduction?

_____

_____

_____

2. What is a galvanic cell?

_____

_____

3. What is the relationship between cell potential and Gibb's free energy?

_____

_____

4. What is a battery?

_____

_____

_____

5. What is electrolysis?

_____

_____

_____

6. How do we determine the mass of product formed during an electrolysis process?

_____

_____

_____

NUCLEAR CHEMISTRY

As you work through your AP Focus Review Guide, keep this chapter's Big Ideas in mind:

BIG IDEA 4 The rate of nucelar decay is measured in half-lives, which is the time requried for a reactant to reach one-half of its initial concentration.

Section 19.1 The Nature of Nuclear Reactions, *pp. 863–865*

Extending Knowledge

This section takes the AP Essential Knowledge you have learned further, and may provide illustrative examples useful for the AP Exam.

REVIEW IT

Complete the following table detailing the particles used in the balancing of nuclear reactions:

Name	Symbol	Mass Number	Atomic Number
proton			
neutron			
electron			
positron			
α particle			

Student Edition p. 864

What is another term for an electron in the context of nuclear reactions?

Section 19.2 Nuclear Stability, *pp. 865–870*

Extending Knowledge

This section takes the AP Essential Knowledge you have learned further, and may provide illustrative examples useful for the AP Exam.

USE IT

Student Edition pp. 869–870

If the sun produces energy at a rate of 5×10^{26} J/s, what is the corresponding mass loss in kg/s?

Section 19.2 Nuclear Stability (continued)

Student Edition
pp. 865–866

Predict which nuclei in each pair is more stable.

$^{6}_{3}\text{Li}$ or $^{9}_{3}\text{Li}$

_____

_____

$^{23}_{11}\text{Na}$ or $^{25}_{11}\text{Na}$

_____

_____

Section 19.3 Natural Radioactivity, *pp. 870–874*

Essential Knowledge Covered:
4.A.3

SUMMARIZE IT

EK 4.A.3

In the thorium decay sequence, thorium-232 loses a total of 6 α particles and 4 β particles in 10 steps. What is the final isotope produced?

Section 19.4   Nuclear Transmutation, *pp. 874–877*

Extending Knowledge

This section takes the AP Essential Knowledge you have learned further, and may provide illustrative examples useful for the AP Exam.

USE IT

Student Edition p. 875

Convert the following reaction into shorthand format.

$^{26}_{12}Mg + ^{1}_{1}p \rightarrow ^{4}_{2}\alpha + ^{23}_{11}Na$

Student Edition p. 876

Write balanced nuclear equation for the following and identify X.

$^{80}_{34}Se$ (d,p) X

Section 19.5   Nuclear Fission, *pp. 877–883*

Extending Knowledge

This section takes the AP Essential Knowledge you have learned further, and may provide illustrative examples useful for the AP Exam.

USE IT

Student Edition p. 883

What is one advantage and one disadvantage of a heavy water reactor in comparison to a light water reactor?

_____

_____

_____

_____

Section 19.6 Nuclear Fusion, *pp. 883–886*

Extending Knowledge

This section takes the AP Essential Knowledge you have learned further, and may provide illustrative examples useful for the AP Exam.

SUMMARIZE IT

Why do heavy elements undergo fission, but light elements undergo fusion? Isn't this a contradiction?

_____

_____

_____

Section 19.7 Uses of Isotopes, *pp. 886–888*

Extending Knowledge

This section takes the AP Essential Knowledge you have learned further, and may provide illustrative examples useful for the AP Exam.

USE IT

Student Edition pp. 886–888

Complete the table that follows, which lists several tracers common in medicine.

Isotope	Half-Life	Uses
^{18}F		
^{43}K		
^{47}Ca		
^{60}Co		
^{99}Tc		
^{125}I		

Section 19.8 Biological Effects of Radiation, *pp. 888–890*

Extending Knowledge

This section takes the AP Essential Knowledge you have learned further, and may provide illustrative examples useful for the AP Exam.

USE IT

Student Edition p. 888

What properties of gamma rays afford them such strong penetrative power?

_____

_____

Student Edition p. 889

Provide an example and chemical explanation for how nuclear radiation affects a biological system.

_____

_____

_____

AP **Reviewing the Essential Questions**

These questions were posed in the *Chemistry* chapter opener (page 862). Answer them using the knowledge you've gained from this chapter.

1. What is radioactivity?

_____

_____

_____

2. What is radiocarbon dating?

_____

_____

_____

As you work through your AP Focus Review Guide, keep this chapter's Big Ideas in mind:

AP A LOOK AHEAD

BIG IDEA 1 The chemical identities of metals are determined by the arrangement of their atoms.

BIG IDEA 2 The properties of ores and minerals are determined by the bonds of their composite elements.

BIG IDEA 3 The process of refining metals and metals combined with non-metals in ores will look at the re-organization of atoms and transferring of electrons as they are separated or extracted.

Section 21.1 Occurrence of Metals, *pp. 931–932*

Extending Knowledge

This section takes the AP Essential Knowledge you have learned further, and may provide illustrative examples useful for the AP Exam.

REVIEW IT

Student Edition p. 931   Outline the difference between a mineral and an ore.

_____

_____

_____

Student Edition p. 932   Where in the periodic table are metals located?

_____

_____

Student Edition p. 932   Based on their location, what property of metals makes finding metal ores more likely than finding pure metals in our environment?

_____

_____

_____

Extending Knowledge
This section takes the AP Essential Knowledge you have learned further, and may provide illustrative examples useful for the AP Exam.

REVIEW IT
Using the word bank on the left, **fill in the blanks:**

Word Bank

alloy
compounds
metallurgy

_____ is the science and technology of separating metals from their ores and of compounding alloys. A(n) _____ is a solid solution of two or more metals, or a metal and other metal _____ involving non-metals.

USE IT

Student Edition p. 934

Match the terms on the left with their definitions on the right.

A. amalgam	1. the waste materials, usually clay and silicate in the preliminary treatment of an ore
B. slag	2. an alloy of mercury with another metal or metals
C. gangue	3. a mixture of calcium silicate and calcium aluminate that remains molten at the furnace temperature

Section 21.3   Band Theory of Electrical Conductivity, *pp. 939–941*

Essential Knowledge Covered:
2.D.2, 2.D.3

REVIEW IT

EK 2.D.2, 2.D.3
Student Edition p. 940

Explain why an insulator such as wood does not conduct electricity.

_____

_____

_____

_____

USE IT

EK 2.D.2, 2.D.3
Student Edition p. 940

Explain the process of *doping* in a semiconductor.

_____

_____

_____

_____

Section 21.3 Band Theory of Electrical Conductivity (continued)

SUMMARIZE IT

EK 2.D.3
Student Edition
pp. 940–941

Outline the difference between a *p*-type and an *n*-type semiconductor. Include a sketch of the atomic array in each type.

Section 21.4 Periodic Trends in Metallic Properties, *pp. 941–942*

Essential Knowledge Covered:
1.C.1

REVIEW IT

EK 1.C.1
Student Edition p. 941

State the trend in metallic character on the periodic table.

EK 1.C.1
Student Edition p. 941

Due to the low electronegativities of metals, which type of ions do they form and which type of oxidation numbers do they have?

Section 21.5   The Alkali Metals, *pp. 942–946*

Essential Knowledge Covered:
1.C.1

REVIEW IT

EK 1.C.1
Student Edition p. 942

What are some of the physical properties associated with the alkali metals?

_____

_____

_____

EK 1.C.1
Student Edition p. 944

Why are sodium and potassium important to living organisms?

_____

_____

_____

Section 21.6   The Alkaline Earth Metals, *pp. 946–948*

Essential Knowledge Covered:
1.C.1, 3.C.3

REVIEW IT

Using the word bank on the left, **fill in the blanks.**

Word Bank

hydroxides
less
more
oxidation

Alkaline earth metals are _____ reactive than alkali metals. In general, they all have a(n) _____ number of +2. The _____ of these elements are extremely strong bases. They tend to react slowly with water to form these bases.

EK 1.C.1
Student Edition
pp. 947–948

What will form in the following reactions?

$Mg(s) + H_2O(l) \rightarrow$

_____

$Ca(s) + H_2O(l) \rightarrow$

_____

USE IT

EK 3.C.3
Student Edition p. 949

What is the overall reaction for the Hall process?

_____

Essential Knowledge Covered:
3.C.3

EK 3.C.3
Student Edition p. 948

USE IT

Outline the process for producing aluminum, using the graphic organizer below.

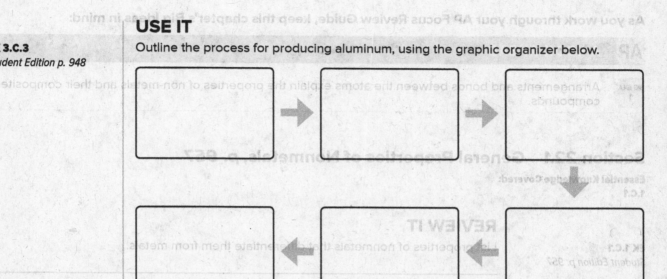

AP Reviewing the Essential Questions

These questions were posed in the *Chemistry* chapter opener (page 930). Answer them using the knowledge you've gained from this chapter.

1. What is the band theory of bonding in metals? Why are metals good conductors?

_____

_____

_____

2. What is a semiconductor?

_____

_____

3. What are some periodic properties of alkali metals and alkaline earths?

_____

_____

4. How is aluminum produced?

_____

_____

NONMETALLIC ELEMENTS AND THEIR COMPOUNDS

As you work through your AP Focus Review Guide, keep this chapter's Big Ideas in mind:

AP   A LOOK AHEAD

BIG IDEA 1   Arrangements and bonds between the atoms explain the properties of non-metals and their composite compounds.

Section 22.1   General Properties of Nonmetals, *p. 957*

Essential Knowledge Covered:
1.C.1

EK 1.C.1
Student Edition p. 957

REVIEW IT

List properties of nonmetals that differentiate them from metals.

_____

_____

EK 1.C.1
Student Edition p. 957

Where are the nonmetals located on the periodic table?

_____

_____

Section 22.2   Hydrogen, *pp. 958–963*

Prerequisite Knowledge

This section covers concepts you should know from other science courses. Review it to master the AP Essential Knowledge and prepare for the AP Exam.

REVIEW IT

Using the Word Bank on the left, **fill in the blanks.**

Word Bank

acid
abundant
covalent
electron
hydrides
ionic
interstitial
proton
water

Hydrogen is the most _____ element in the universe, accounting for approximately 70% of the total mass in the universe. It is an element made up of one _____ and one _____. The mixture of carbon monoxide and hydrogen gas is more commonly known as _____ gas. Small scale preparation of hydrogen gas can be achieved by the reaction of a metal such as zinc with a(n) _____.

Binary _____ are compounds containing hydrogen and another element, either a metal or nonmetal. There are three types of these compounds, _____, _____ and _____.

Section 22.3   Carbon, *pp. 963–966*

Prerequisite Knowledge

This section covers concepts you should know from other science courses. Review it to master the AP Essential Knowledge and prepare for the AP Exam.

REVIEW IT

Using the word bank on the left, **fill in the blanks.**

Word Bank

catenation
carbides
coal
diamond
dry ice
graphite
natural gas
petroleum

While only approximately 0.09% of the mass of the Earth's crust is carbon, it is an essential component of living matter. It is found naturally in the elemental form of _____ and _____. Most carbon in nature is in the form of _____, _____, and _____.

Carbon also has the ability to form long chain and ring molecules. The process of forming long chain molecules is called _____. When carbon forms compounds with metals, the resulting molecules are referred to as _____.

When carbon dioxide is in its solid form, it is referred to as _____.

Section 22.4   Nitrogen and Phosphorus, *pp. 967–974*

Prerequisite Knowledge

This section covers concepts you should know from other science courses. Review it to master the AP Essential Knowledge and prepare for the AP Exam.

REVIEW IT

Student Edition p. 967

What is the approximate percentage (by volume) of the air that is nitrogen?

_____

Student Edition p. 968

Ammonia can be prepared in the lab by reacting ammonium chloride with sodium hydroxide. Write this reaction.

_____

Student Edition p. 970

White phosphorus exists as $P_4(s)$ and is not only highly toxic, it is also extremely flammable. What will be the molecule that forms when this material burns in the presence of oxygen? Write the balanced equation for this process.

_____

Section 22.5   Oxygen and Sulfur, *pp. 975–982*

REVIEW IT

Student Edition p. 975

Draw chemical structure diagrams for the two allotropes of oxygen.

Student Edition p. 975

Write the Lewis structures for the oxide, peroxide and superoxide ions.

USE IT

Student Edition p. 976

Why is hydrogen peroxide usually placed in a container that does not allow exposure to sunlight?

What chemical reaction will occur if sunlight is allowed to react with the hydrogen peroxide?

Section 22.6   The Halogens, *pp. 982–989*

Essential Knowledge Covered:
1.C.1

REVIEW IT

EK 1.C.1
Student Edition p. 985

Due to its high electronegativity, fluorine only has two possible oxidation numbers. What are these oxidation numbers and when do they occur?

USE IT

EK 1.C.1
Student Edition p. 982

What valence electron configuration do all of the halogens have in common?

_____

_____

SUMMARIZE IT

EK 1.C.1

Fluorine is prepared by the electrolysis of liquid hydrogen fluoride.

Write the half reactions and overall reaction to this process.

Draw a representative electrolytic cell for this process.

These questions were posed in the *Chemistry* chapter opener (page 956). Answer them using the knowledge you've gained from this chapter.

1. What are some periodic properties of nonmetals? How does atomic radius and the first ionization potential change from fluorine to iodine?

_____

_____

_____

_____

As you work through your AP Focus Review Guide, keep this chapter's Big Ideas in mind:

AP ▸ A LOOK AHEAD

BIG IDEA 1 The transition metal compounds have structural features that make them very useful in industrial and commercial processes.

Section 23.1 Properties of the Transition Metals, *pp. 995–998*

Essential Knowledge Covered:
1.C.1

SUMMARIZE IT

EK 1.C.1

List at least two differences that exist between metals (such as tin or lead) and the transition metals.

_____

_____

_____

Section 23.2 Chemistry of Iron and Copper, *pp. 998–1000*

> ### Extending Knowledge
> This section takes the AP Essential Knowledge you have learned further, and may provide illustrative examples useful for the AP Exam.

USE IT

Student Edition pp. 998–1000

Give the valence electron configuration for both iron and copper, and indicate the electrons that are ionized to form each of the two common oxidation states as given in the table.

Metal	Valence configuration	Electrons ionized to form +2	Electrons ionized to form +3
iron			
copper			

Which element in the above table would you expect silver (Ag) to behave more similarly to? Why?

_____

_____

_____

Extending Knowledge

This section takes the AP Essential Knowledge you have learned further, and may provide illustrative examples useful for the AP Exam.

USE IT

Student Edition p. 1003

Name the following coordination compounds according to the rules on page 1003 of your textbook.

$[Co(NH_3)_4Cl_2]Cl$ _____

$Na_3[AlF_6]$ _____

$[Co(en)_2Cl_2]NO_3$ _____

$[Pt(NH_3)_4BrCl]Cl_2$ _____

$[Co(NH_3)_6][FeCl_4]_3$ _____

Section 23.4   Structure of Coordination Compounds, *pp. 1005–1008*

Extending Knowledge

This section takes the AP Essential Knowledge you have learned further, and may provide illustrative examples useful for the AP Exam.

SUMMARIZE IT

How are structures A and B related? Explain your reasoning.

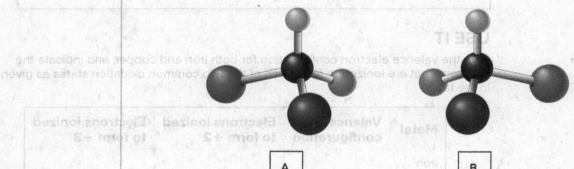

_____

_____

_____

_____

Section 23.5　Bonding in Coordination Compounds: Crystal Field Theory, *pp. 1009–1014*

Extending Knowledge

This section takes the AP Essential Knowledge you have learned further, and may provide illustrative examples useful for the AP Exam.

USE IT

Student Edition p. 1011

How does crystal field theory account for coloration in coordination complexes?

_____

_____

_____

_____

_____

Section 23.6　Reactions of Coordination Compounds, *pp. 1015–1016*

Extending Knowledge

This section takes the AP Essential Knowledge you have learned further, and may provide illustrative examples useful for the AP Exam.

SUMMARIZE IT

Why does the color of copper(II) chloride solution change upon dilution with water? Is the rate of this color change likely to change with a change in temperature?

_____

_____

_____

_____

Section 23.7 Applications of Coordination Compounds, *pp. 1016–1020*

Student Edition pp. 1018–1020

USE IT

How does the structure of a coordination compound such as cyanide or carbon monoxide affect the toxicity of these compounds to living organisms?

_____

_____

_____

_____

SUMMARIZE IT

Based on your answer to the above, would the iron-cyanide complex be considered inert or labile to an oxygen ligand?

_____

_____

_____

These questions were posed in the *Chemistry* chapter opener (page 994). Answer them using the knowledge you've gained from this chapter.

1. What are some periodic properties of transition metals?

_____

_____

_____

_____

As you work through your AP Focus Review Guide, keep this chapter's Big Ideas in mind:

AP   A LOOK AHEAD

BIG IDEA 2   The fundamental building blocks of matter, atoms, link together to form extremely complex molecules in organic chemistry.

BIG IDEA 5   The attractive forces between positive and negative charges in organic molecules make life on Earth possible.

Section 25.1   Properties of Polymers, *p. 1059*

> ### Extending Knowledge
>
> This section takes the AP Essential Knowledge you have learned further, and may provide illustrative examples useful for the AP Exam.

Student Edition p. 1059

REVIEW IT

Define the term *polymer*.

_____

_____

Student Edition p. 1059

Polymers

cellulose
nucleic acids
nylon
poly (methyl
  methacrylate)
polyethylene
proteins
rubber
wood

Macromolecules are generally separated into naturally occurring polymers and synthetic polymers. Place the polymers given in the list in the left column in the proper column in the table.

Synthetic	Naturally Occurring

Extending Knowledge

This section takes the AP Essential Knowledge you have learned further, and may provide illustrative examples useful for the AP Exam.

REVIEW IT

Student Edition p. 1059

Define the term *monomer*.

_____

Student Edition p. 1059

What are the two types of isomers?

_____

Student Edition pp. 1059, 1063

What are the two main types of reactions that form synthetic polymers from monomer subunits? How do they occur?

_____

_____

_____

_____

_____

Section 25.3   Proteins, *pp. 1065–1073*

Essential Knowledge Covered:
2.B.2, 5.D.3

REVIEW IT

EK 2.B.2, 5.D.3
Student Edition p. 1065

What two groups do all amino acids have in common? Draw a representative diagram of each group.

_____

_____

_____

_____

_____

_____

EK 2.B.2, 5.D.3
Student Edition p. 1065

How do amino acids synthesize into proteins?

_____

_____

SUMMARIZE IT

EK 2.B.2, 5.D.3

Describe the four levels of protein organization: primary, secondary, tertiary and quaternary structures.

_____

_____

_____

_____

_____

_____

_____

_____

_____

Essential Knowledge Covered:
2.B.2, 5.D.3

EK 2.B.2
Student Edition p. 1073

REVIEW IT

Define the term *nucleic acids*. What role do they play in living organisms?

_____

_____

EK 2.B.2, 5.D.3

SUMMARIZE IT

Outline the three rules proposed by Erwin Chargaff in the 1940's associated with DNA molecules.

1. _____

_____

2. _____

_____

3. _____

_____

EK 2.B.2, 5.D.3

What are the three major differences between DNA and RNA?

1. _____

_____

2. _____

_____

3. _____

_____

These questions were posed in the *Chemistry* chapter opener (page 1058). Answer them using the knowledge you've gained from this chapter.

1. What types of intermolecular forces occur in proteins and nucleic acids?

_____

_____

_____

2. How does the structure of a biological compound influence its function?

_____

_____

_____